Concrete Admixtures

Concrete Admixtures: Use and Applications

Edited by
M. R. Rixom

on behalf of
The Cement Admixtures Association

THE CONSTRUCTION PRESS
LANCASTER   LONDON   NEW YORK

The Construction Press Ltd.,
Lancaster, England.

A subsidiary company of Longman Group Ltd., London.
Associated companies, branches and representatives
throughout the world.

Published in the United States of America by
Longman Inc. New York.

© Cement Admixtures Association Limited

All rights reserved. No part of this publication may be
reproduced, stored in a retrieval system, or transmitted
in any form or by any means, electronic, mechanical,
photocopying, recording, or otherwise, without the prior
permission of the Copyright owner.

First published 1977.

ISBN 0 904406 32 6

Printed by The Scolar Press Limited, Ilkley, England.

Contents

Foreword

1 An Introduction to and a Classification of Cement Admixtures   9
  P. C. Hewlett

2 Water Reducing Admixtures for Concrete   23
  M. R. Rixom

3 Air Entraining Admixtures for Concrete   37
  K. Bennett

4 Pumping Aids for Concrete   49
  A. S. Flatau and L. H. Mc Currich

5 Integral Waterproofers for Concrete   55
  P. C. Hewlett, R. W. Edmeades and R. L. Holdsworth

6 Mortar Plasticisers   67
  H. E. Akerman

Appendix
    Concrete Admixtures – Data Sheets   73

Foreword

The Cement Admixtures Association is a trade association formed to promote the wider acceptance and use of admixtures in concrete and mortars. In this respect over recent years it has successfully organised a number of internationally attended conferences and seminars and has played a major role in the area of education in concrete technology by providing speakers and lecturers for courses organised principally in the U.K.

As a further step towards widening the knowledge of materials produced by member companies, the Technical Committee of the Association, mainly under the consecutive Chairmanships of Dr P.C. Hewlett and Mr M.R. Rixom, have produced this book which is based on a collection of papers presented by Association members. It is intended to provide general background information on the major types of admixtures used in the production of concrete and mortars, and to assist all those engaged in the manufacture, specification and use of concrete to understand the benefits that are to be obtained from their use. When it is considered that in 1975 some 12-15M m^3 of concrete were produced in the U.K. containing some type of chemical admixture then the importance will be realised of ensuring that those people engaged in the production or handling of concrete have a sound understanding of the materials being utilised.

J.D. Lincoln
Chairman, Cement Admixtures Association
April, 1977

1

An Introduction to and a Classification of Cement Admixtures

P C Hewlett

INTRODUCTION

Definition of admixtures

Admixtures are materials that are added to concrete at some stage in its making to give to the concrete new properties either when fluid or plastic and/or in the set or cured condition. Admixtures differ from additives, which are materials added to the cement during its manufacture either as an aid to production, such as a grinding aid, or when the cement is to be used to make concrete with special properties, such as certain waterproof concretes. In this respect Gypsum is an additive.

General attitude towards admixtures

Concrete outside of the laboratory and away from the committees responsible for specifications and codes of practice is a much abused material, and the general attitude – until recent years – towards admixtures was one of disbelief or disinterest. Perhaps it is because concrete has been used for so long and has become commonplace that we do not regard it as a material in the true sense. This attitude coupled with a degree of 'Witch Doctor' chemistry surrounding admixtures has led to a slow and reluctant acceptance of these materials. In contrast it is interesting to note that admixtures in concrete have been more widely accepted in the U.S.A. and on the Continent – particularly in Germany – than in the United Kingdom. However, with the present trend in cement price it is probably timely that the users and manufacturers of concrete should take a keen and informed interest in the part that admixtures could play in making the best use of the cement paid for.

CLASSIFICATION OF ADMIXTURES

What then are admixtures? What is available? How do they work and what are the benefits as well as the shortcomings?

There are, for convenience, eight main categories of admixtures and five sub-groups and these are shown in Table 1/1. These admixtures are by no means used to an equivalent extent. At the top of the tonnage league we have the air-entraining and water-reducing materials and at the bottom the pumping aids and the miscellaneous group. Obviously there will be regional variations. For instance, cold areas may well favour accelerators and air-entraining agents and by contrast warmer zones favour retarders. Areas noted for poor or harsh aggregates might favour plasticisers or workability aids. However, each category will be dealt with in turn ignoring regional variations.

Table 1/1 Classification of admixtures

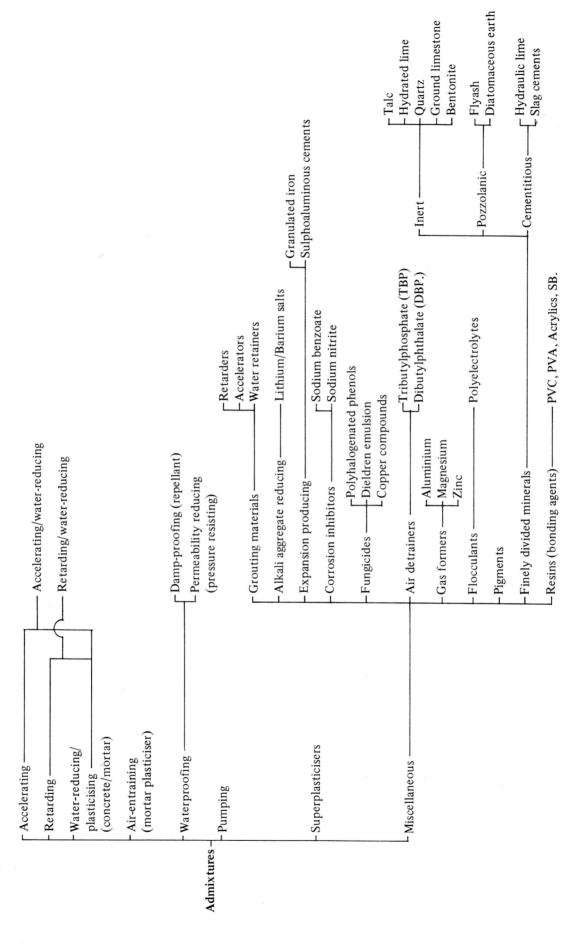

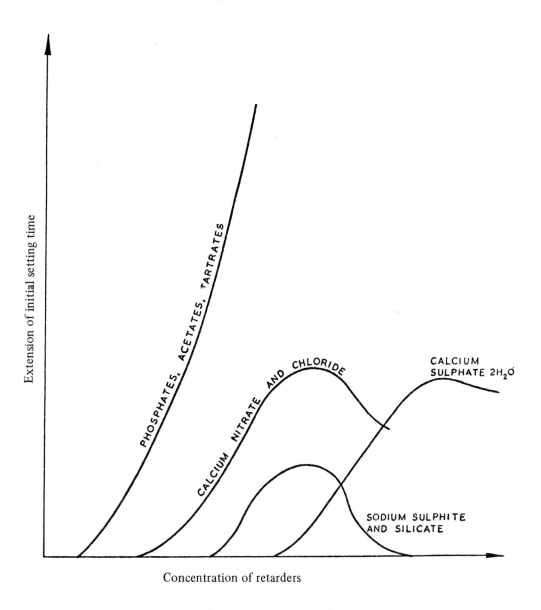

Figure 1/1 Effect of various admixtures on initial setting time.

Accelerating admixtures (category C1 admixture chart)

These materials shorten the setting time of cement and/or increase the rate of strength build up. They help in maintaining schedules during cold weather working, but in addition allow early removal of formwork and a general reduction in contractors' time-tables.

In the majority of products in this category the accelerating chemical is calcium chloride although other materials such as triethanolamine, calcium nitrite, calcium formate, lithium oxalate and certain aluminates also have an accelerating effect. A few chemicals having plasticising action are sometimes classed as accelerators although they themselves do not partake in the speeding up of the cement hydration reaction. Calcium chloride has been criticised severely over the last few years as a result of reinforcement corrosion associated with chloride action, and recently its use has been banned in the U.K. for any concrete containing embedded metal. It remains an acceptable and highly effective admixture for unreinforced concrete.

Because of the corrosion problems associated with calcium chloride, chloride free accelerators based on calcium formate are becoming widely used. The mechanism by which accelerators work is not completely known but it appears that both calcium chloride and calcium formate tend to

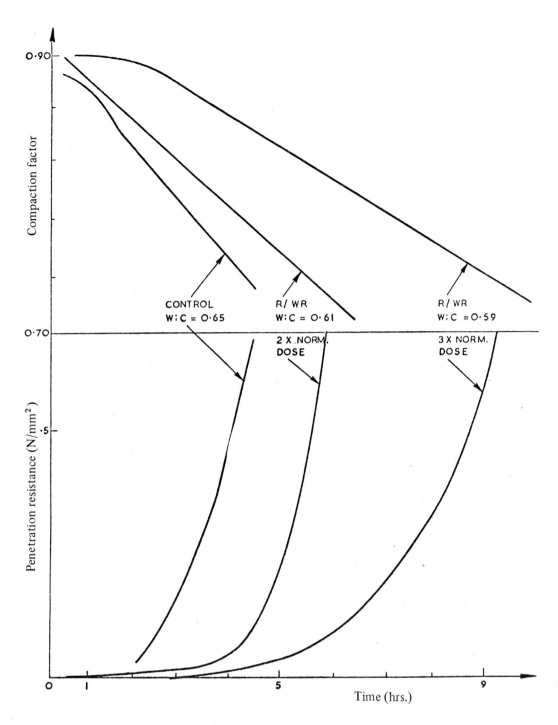

Figure 1/2 Effect of retarding/water-reducing admixtures on retention of workability.

increase the rate of hydration of the different chemical constituents in the cement. Mixes containing accelerators must still be protected from frost during the early stages of setting (until strength has reached approximately 5 N/mm²).

Retarders (category C2 – admixture chart)

Retarders slow down the rate of setting of cement and as with retarding/water-reducing admixtures they assist in hot weather concreting and in the casting and consolidating of large numbers of pours without the formation of cold joints. They do this by extending the vibration time. However, once the reaction has got through the setting stage the hydration continues at the normal rate and sometimes greater. Retarders are used to ostensibly offset heat build-up by presumably slowing down the hydration reaction and allowing any heat generated to dissipate. Whilst used quite effectively for making large pours a practical proposition there is little evidence to show that the total heat output or rate of heat output is in fact markedly altered by the addition of a retarder.

Materials comprising retarders are in the main:—

1. Unrefined lignosulphonates containing sugars, which are, of course, the component responsible for retardation.
2. Modifications and derivatives of the first group.
3. Hydroxycarboxylic acids and their salts, for instance sodium tartrate.
4. Modifications to the above class.
5. Carbohydrates including sugars.
6. Heptonates which are related to the sugars and starches.

Types 1 to 4 and 6 also have water reducing properties; type 5 does not. It is thought that retarding admixtures are absorbed on to the C_3A phase in cement forming a film around the cement grains and preventing or reducing the reaction with water. After a while this film breaks down and normal hydration proceeds. This is a very simple picture and there is reason to believe that retarders also interact with the C_3S phase since retardation can be extended to a period of many days which cannot be accounted for by the C_3A/water reaction alone.

It is interesting to note that some well known accelerators can in turn act as retarders depending on concentration, particularly when used at low concentrations. Some of these materials are compared in relation to the conventional retarder in Figure 1/1. The retention of workability using a retarder is shown in Figure 1/2.

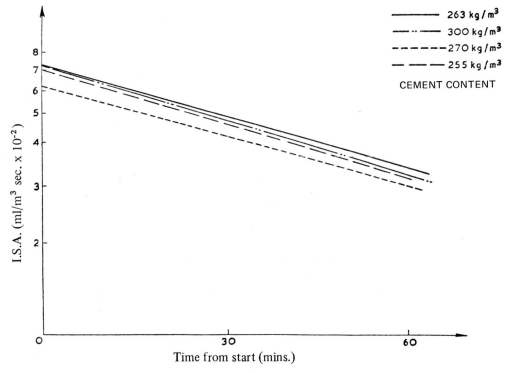

Figure 1/3 Initial surface absorption results with water-reducing/plasticising admixtures.

Water-reducing or plasticising admixtures (category C3 – admixture chart)

The addition of a plasticiser allows greater workability to be achieved for a given water:cement ratio or alternatively retains the workability or consistency whilst reducing the water content.

The latter use results in denser and stronger concrete for a given cement content. This property has been and can be used to effect cement savings whilst maintaining strength. The use in this context has raised questions about the long term durability of such concretes but in the cement content range 300 kg/m^3 or thereabouts those properties controlling durability, such as permeability and porosity, appear unaffected by as much as 10% cement reduction in the presence of a plasticiser, as can be seen from the initial surface absorption results in Figure 1/3.

How they work

Let us consider for a moment how water-reducing agents actually function. As a group of materials water-reducing agents are characterised by having detergent-like properties. These we call surface activity, and the materials surface active agents. They carry an unbalanced charge of electricity and if put into water tend to migrate to the surface with the electrically charged or active end sticking into the water, whilst the 'tail' is out in the air (Figure 1/4).

If we now put a surface active agent into a suspension of cement particles in water two things happen:

1. The surface active agent's 'tail' is adsorbed on to the surface of the cement with the negative charge protruding into the water. As a result the cement particles do not collect together and therefore more surface area is available for reaction with the water. At the same time water that may have been trapped inside a cement particle floc is released. The combined effects, shown in Figure 1/5, improve the workability and mobility of the cement mix.

2. Entrapped air is also more readily removed since orientation of the surface active agent prevents the air bubble from attaching to cement particles, as shown in Figure 1/6.

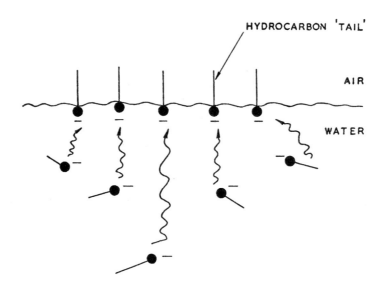

Figure 1/4 Migration of water-reducing agents to surface of water.

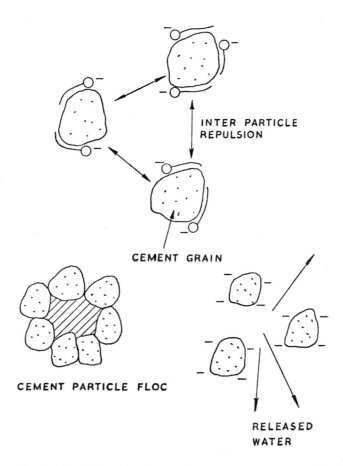

Figure 1/5 Effect of surface-active agent on cement particle floc.

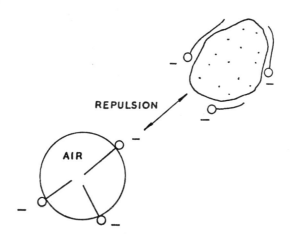

Figure 1/6 Repulsion of air bubble by surface-active agent.

Plasticising water-reducers, due to their ability to migrate to the water/air boundary, reduce the surface tension at the boundary and can form very stable air bubbles. This is the principle behind air-entraining agents.

However, in plasticisers this tendency can be offset by introducing an air-detraining agent.

Why they are used　　If concrete is being made to a given workability, say slump value, then the addition of a water-reducing agent gives rise to the following possibilities:

15

1. Concretes having greater workability may be made without the need for more water and so strength losses are not encountered.
2. By maintaining workability, but at a lower water content, concrete strengths may be increased without the need for further cement addition.
3. Whilst maintaining the water:cement ratio and workability concrete can be made to a given strength specification at lower cement contents than would otherwise be required.

Table 1/2 reports results for concretes for each of the possibilities given above.

Water-reducing agents can give benefit in concrete containing harsh and/or poorly graded aggregates and for concrete being placed under difficult conditions such as using a Tremie.

In order to recoup the benefit of using a water-reducing agent care must be taken in controlling the air content of the mix. Most water-reducing agents entrain air due to their surfactant properties. This can be offset by including in this type of admixture an air detraining agent such as tributylphosphate which is reflected in the amount of air entrained and the water reduction obtained (as shown in Table 1/3).

Table 1/2 Purposes achieved by use of water-reducing agents.

Test	Cement content (kg/m^3)	Dosage rate (litres/50 kg)	*W:C ratio	Slump (mm)	†UCS (N/mm^2) 7 day	28 day
Control	300	0	0.62	50	25.0	37.0
Workability increase	300	0.14	0.62	100	26.0	38.0
Strength increase	300	0.14	0.56	50	34.0	46.0
Cement saving	270	0.14	0.62	50	25.5	37.5

*Water:cement
†Unconfined compressive strength

Table 1/3 Use of an air-detraining agent with a water-reducing agent.

Admixture	Dosage (%)	Slump (mm)	Air content (%)	Water reduction (%)
Control	0	89	2.0	—
Lignosulphate water-reducing agent (unmodified)	0.16	89	3.8	11.3
	0.26	114	5.0	14.2
	0.37	114	6.5	17.9
Lignosulphate water-reducing agent (modified)	0.16	102	2.5	5.8
	0.26	89	3.3	9.9
	0.37	102	3.8	12.8

Types of materials

1. Lignosulphonate salts

These are naturally occurring materials which are obtained by pulping wood using lime and sulphurous acid mixture. They are usually described as calcium lignosulphonates and may or may not contain a proportion of sugars. If sugars are present retardation may occur; alternatively a refined, sugar-reduced product may be used which avoids excessive retardation.

2. Polyhydroxy compounds (carboxylic acid based)

These materials are similar in function to the lignosulphonates and may be sensitive to over-dosing leading to excessive retardation.

3. Superplasticisers

A newer range of plasticising admixtures is now available based on synthesized condensates. Such materials as a group could be specified as formaldehyde derivatives such as melamine-formaldehyde and naphthalene sulphonate-formaldehyde. These materials have quite remarkable plasticising action, producing slumps in excess of 200 mm with no increase in water content.

This plasticising effect is being used both on the Continent and to some degree in the U.S.A. and Japan for projecting the concept of 'fluid' or 'soupy' concrete. The disadvantage of these materials is their high basic cost, although each product should be judged on its cost-in-use. It is worth mentioning that chemicals capable of imparting high fluidity to a concrete should also offset bleed and segregation in order to be totally effective.

Accelerating/water-reducing admixtures (category C4 — admixture chart)

Almost without exception accelerating/water-reducing materials are mixtures of clacium chloride (which is the accelerating component) blended with a plasticiser water-reducing agent.

All accelerating/water reducing admixtures increase the rate of hydration of the cement resulting in a more rapid increase in the build-up of the strengths, which is particularly useful at low ambient temperatures, as shown in Tables 1/4 and 1/5.

This is achieved by the calcium chloride reacting mainly with the C_3A/C_2S phases in the cement (these phases are responsible for the setting characteristics of the cement and later strength development respectively).

The use of such admixtures aids cold weather working and lessens the susceptibility to frost damage.

Table 1/4 Effect of accelerating/water-reducing admixtures at low ambient temperatures.

Temperature of casting or curing	Mix (300 kg/m³) slump – 50mm	W:C ratio	UCS (N/mm²)			
			1 day	3 day	7 day	28 day
2°C	Control	0.65	–	3.6	12.8	26.1
	Admixture	0.60	–	7.2	19.3	32.0
5°C	Control	0.65	–	5.4	16.7	30.9
	Admixture	0.60	–	8.0	21.0	35.9
18°C	Control	0.65	5.6	15.5	28.1	38.6
	Admixture	0.60	7.9	19.2	32.9	41.6

Table 1/5 Effect of accelerating/water-reducing admixture on setting time

Mix	Setting time (mins)	
	Initial	Final
Control	140	200
Admixture (normal dose)	75	115
" (double dose)	65	95

Table 1/6 Effect of retarding/water-reducing admixtures on setting time and strength build-up.

Admixture addition (litres/50 kg)	Setting time Proctor needle (hrs.)		W:C ratio	UCS (N/mm^2)		
	Initial	Final		3 day	7 day	28 day
0	4.5	9	0.68	20.3	28.0	37.0
0.14	8	13	0.61	28.0	36.5	46.8
0.21	11.5	16	0.58	29.6	40.1	49.7
0.28	16	21	0.58	30.1	41.6	54.1

Table 1/7 Effect of retarding/water-reducing admixtures at high ambient temperatures.

Test	Ambient temperature (°C)	Concrete slump (mm)					
		Time (hrs.)					
		0	1	2	3	4	5
Control (no admixture)	20	127	89	76	57	38	32
+ Admixture	20	127	127	114	102	70	57
Control (no admixture)	43	114	57	7	—	—	—
+ Admixture	43	127	70	25	19	13	—

It also allows

(i) earlier release and re-use of formwork and moulds.

(ii) earlier handling of precast concrete.

(iii) earlier finishing of concrete floors and screeds.

The above advantages result in savings in time, labour and materials.

It is interesting to note that plasticising accelerators are of particular use in flooring and screeding applications where time between initial and final trowelling can be reduced by as much as 50%.

Retarding/water-reducing admixtures (category C5 – admixture chart)

These are usually mixtures of conventional water-reducing agents plus sugars or hydroxycarboxylic acids or their salts.

Both the setting time and the rate of strength build-up are affected by these materials, as shown in Table 1/6.

The main use of retarding plasticisers is to seemingly reduce the rate of heat generation from concrete so lessening the chance of thermal cracking and easing the making and placing of large pours. There is some doubt that retarders or retarding plasticisers really alter the total heat output or, for that matter, the rate of heat generation. Certainly the strength/time figures contradict the simple explanation and more fundamental work is needed here. However, retarding plasticisers are very effective in maintaining workability for long periods and help in offsetting the effect of high ambient temperatures, which is illustrated in Table 1/7.

Air-entraining admixtures (category C6 – admixture chart)

Concrete containing small air bubbles (0.05 mm/1.25 mm) spaced at gaps less than or equal to 0.4 mm (the bubble size is about one thousand times that of the cement paste capillary pores) evenly distributed throughout its bulk is demonstrably more durable to freeze/thaw action than normal concrete (Figure 1/7.)

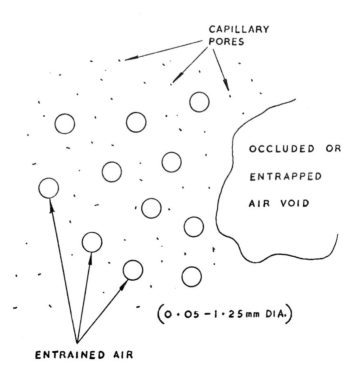

Figure 1/7 Even distribution of entrained air.

Destructive pressures up to 8 N/mm² can be generated as a result of freeze/thaw action. It also resists better the destructive action of de-icing salts. Air-entrained concrete whilst in the plastic state has improved workability. It can be handled and placed with less segregation and bleeding. Of course, one never gets something for nothing and air entrainment probably results in some strength loss. For instance 10 - 50% in U.C.S. is commonplace at an air entrainment of 10 - 13% by volume of the mortar fraction of the concrete, but can at least partially be compensated for by mix design.

Materials comprising this group are:—

1. Natural wood resins and their soaps (e.g. Vinsol resin).
2. Certain fats and oils (tallow, olive oil and stearic acid).
3. Lignosulphonates.

All air-entraining agents are surface active and reduce the surface tension at the water/air interface. This means that minute bubbles that form remain as stable bubbles and do not collapse. Dosage control is important in order to entrain the prescribed amount of air. There are interesting new materials based on sulphonated naphthalenes that reach a maximum level of air entrainment so that over-dosing presents no hazard. These materials are being investigated.

Waterproofers (category M2 – admixture chart)

This class of materials, perhaps more than any other admixture, is difficult to vindicate. Firstly, engineers quite rightly maintain that good concrete is waterproof and secondly, to prove that an admixture enhances waterproofness is very difficult indeed. However, in practice much concrete, despite what it should and could be, is often porous and in this state will absorb and transmit moisture and, in severe cases, water. So what can be done about it? Firstly, we should distinguish between water-repelling admixtures and those that prevent or retard the flow of water through mass concrete. The repellents line the pores so making them to some extent hydrophobic, thus increasing the contact angle between the water and the concrete and preventing the passage of water at low pressures — but not the passage of vapour. Waterproofers or permeability reducers on the other hand act by blocking the pores and will to some extent prevent water flow at moderate/medium pressure gradients. Materials within this class are legion and some attempt at rationalisation would reduce the confusion. A brief attempt to classify waterproofers is given below but is dealt with in detail in 'Integral Waterproofers'.

(a) Permeability reducers

These are fine particulate materials such as whiting, bentonite, limestone, colloidal silica, used together with one of three members of the preceding classes, namely, water reducers, air entrainers and accelerating admixtures.

(b) Damp-proofers

These include the soaps such as sodium and ammonium stearate, butyl stearate and a smattering of petroleum products, such as petroleum jelly and naphthalene.

It is thought that under alkaline conditions and in the presence of excess calcium an insoluble calcium salt is formed as a precipitate which either blocks and/or lines the pores of the concrete.

Objective testing of waterproofers is fraught with problems although several attempts have been made to overcome these. However, at a practical level, incorporation of damp-proofers in concrete often brings the absorption test results within the British Standard and can be readily shown to repel water and prevent capillary absorption. (A better test than the simple water absorption test is the ISA as detailed in British Standard 1881).

Pumping aids (category C7 — admixture chart)

Much concrete is pumped in the U.S.A. and a considerable quantity on the Continent but in the U.K., as yet, very little by comparison, and proportionately even less when such concrete contains a pumping aid. The apparent lack of enthusiasm for pumping aids may be due to the good aggregates generally available, but it is likely that within confined city working and in areas where concreting is to be done with aggregates which are not so good or not readily available, such as those parts of Scotland concerned with North Sea exploitation, pumping may get a boost and this, in turn, may encourage the use of an admixture.

It is true that good mix design — at a price — can produce concrete that is readily pumped but certain chemical admixtures — often at less cost — can produce readily pumpable concrete.

Admixtures which aid pumping impart cohesiveness to the mix so that when a slug of concrete is subjected to a pressure gradient there is little or no separation of the liquid and solid phases and the slug remains intact. Without the admixture the water phase moves relative to the aggregate/sand/cement causing the concrete to 'dry out' and wedge by dilating in the lines. The admixture has a secondary effect of lubricating the walls of the lines so that slug flow is induced rather as toothpaste is extruded from a tube.

Materials in this category are the alginates, cellulose ethers and polyethylene oxides with or without modification. An advantage of the polyethylene oxides is that concrete strengths are often improved by their use whereas other pumping aids tend to entrain air, causing strength loss.

Miscellaneous

As can be seen in Table 1/1, this group comprises a large number of material types.

Expansion producers/gas formers

The difference between expansion producers and gas formers is really one of magnitude of the expansion produced.

Granulated iron in contact with alkaline cement oxidises, the products of oxidisation are less dense than the starting material and a small expansion results. This is useful for offsetting the effects of drying shrinkage.

Gas-forming admixtures on the other hand react with the alkali producing a gas rather like soda in bread baking and the resulting expansion is more marked.

Hydrogen peroxide acts by liberating oxygen.

Metallic aluminium acts by liberating hydrogen.

Activated carbon acts by liberating absorbed air.

Addition levels are very small, particularly for aluminium powder, and so these materials are often premixed with another material such as fine sand, cement or a pozzolan to aid dispensing. Gas-forming admixtures act during the plastic stage of concrete and cannot be used to offset drying shrinkage which occurs after hardening.

Finely divided materials

Within these materials I wish to mention the fly ash product, PFA, which is obtained from the residues of solid-fuel-burning power stations as opposed to blast furnace slag.

PFA is pozzolanic – in other words, it reacts with the lime and cement to produce a product having cementitious properties. Blast furnace slag on the other hand reacts with the lime or calcium present in cement in such a way that it is 'activated' and undergoes hydration in a similar manner to the cement itself.

Resin materials

Bonding agents based on PVA, PVP, Acrylics and Styrene/butadiene are well-known. Less exploited are the epoxide resin emulsions (water dispersibles) that upgrade the mechanical and chemical resistance properties of cement mortars being used as thin screeds (less than ½″ thick). Cost prevents their more general use, but expense at the specifying stage could often be offset against savings in repair work. With the present trend in resin prices, and in some cases reduced availability, Styrene/butadiene latices may well get a boost.

Some of the other materials of the miscellaneous group have been dealt with in the ACI Committee 212 report dated 1971 entitled 'Guide for use of Admixtures in Concrete' which is recommended for further reading.

2

Water Reducing Admixtures for Concrete

M R Rixom

INTRODUCTION

This chapter is concerned with the description and use of water-reducing admixtures or, as they are more commonly known, plasticisers for concrete. This product group of admixtures probably accounts for the largest tonnage value of organic materials used in concrete and has accordingly been the subject of fairly extensive fundamental and applied research work both here and overseas. The chapter, after describing the three types of water-reducing admixture, and outlining the mode of action common to them all, will give a more detailed survey of individual types with regard to the following properties:

1. Chemical type
2. Effect on the properties of plastic concrete:
   a) Set
   b) Air content
   c) Workability
   d) Water reduction
3. Effect on the properties of hardened concrete:
   a) Strength development
   b) Durability
   c) Shrinkage
   d) Creep
4. Applications.

TYPES OF WATER-REDUCING ADMIXTURES

There are three main categories of water-reducing admixtures and this chapter will be restricted to these types:

(i) Normal water-reducing admixture

(ii) Accelerating/water-reducing admixture

(iii) Retarding/water-reducing admixture.

All three types can be utilised to perform the following functions:

a) Their addition to a concrete mix will produce an increase in workability measured by tests such as the slump, compacting factor and VeBe.

Table 2/1 Effects of using water-reducing admixtures.

Category of admixture	Water reduction	Stiffening time		Compressive strength		Cement Admixtures Association chart type
		Time from completion of mixing to reach a resistance to penetration of		Percentage of control mix (minimum)	Age	
		0.5 N/mm^2	3.5 N/mm^2			
Normal water-reducing	At least 5%	Within ± 1 h of control mix	Within ± 1 h of control mix	110	7 & 28 days	C3
Accelerating water-reducing	At least 5%	More than 1 h	At least 1 h less than control mix	125 110	24 h 7 & 28 days	C4
Retarding water-reducing	At least 5%	At least 1 h longer than control mix	—	110	7 & 28 days	C5

b) For a given workability the amount of water, that is the water cement ratio, can be reduced.

c) Cement reductions can be made by increasing the aggregate/cement ratio to effect economies without any detrimental effect on workability or compressive strength in comparison with the original mix at periods up to 28 days.

The differences between the types within the group lie in their effect on the set and strength development characteristics of the concrete into which they are incorporated and British Standard 5075 Part 1 1974 – Water Reducing Admixtures for Concrete – gives the criteria for the three group types (Table 2/1).

In addition, a column has been added to the table to give the product grouping from the Cement Admixtures Association Wall Chart (10).

MODE OF ACTION

It was previously stated that all three types within the water-reducing admixture group possessed the ability to increase the workability or alternatively allow similar workabilities to a control mix at lower water/cement ratios. The reason for this is that all three materials contain either as the sole or major ingredient materials possessing surface activity which are able to deflocculate or disperse cement particles.

Portland cement in its dry state exists in the form of agglomerates of primary particles containing many such particles held together by forces ranging from very weak to approaching chemical bond. During the mixing cycle the agglomerates are to an extent broken down into smaller fragments and hydration occurs at the surface of these fragments, enmeshing other hydrating fragments together with the fine and coarse aggregates to produce a composite material of high strength and impermeability. The mechanism by which all members of this admixture group operate is to deflocculate or to disperse the cement agglomerates into primary particles or at least into much smaller fragments. This deflocculation is believed to be a physico-chemical effect whereby the admixture is first of all adsorbed on to the surface of the hydrating cement and, forming a hydration 'sheath', reduces the interparticle attraction, causing the cement particles to be separated from one another and thereby reducing the inter-particle friction. Figure 2/1 illustrates this.

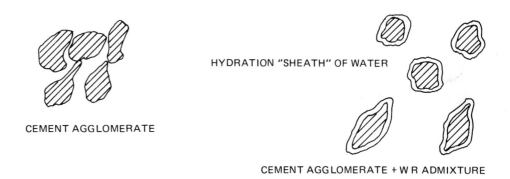

Figure 2/1 **Diagram illustrating deflocculation.**

The results of such a deflocculated system are as follows:—

a) the dispersion of the agglomerates into much smaller primary particles reduces the retention of water which would otherwise be used for imparting workability to the system. In view of this less water is required to achieve a given workability or alternatively more workable concrete can be obtained for a given water content.

b) under conditions of shear or movement within the system there will be less inter-particle or particle/aggregate interaction or friction so that less energy will be required to place concrete containing such a system.

c) the growth of cement hydrates is largely a surface effect and therefore even at constant water/cement ratios a deflocculated system will contain particles spaced at a more uniform distance from one another. Therefore, on continuing hydration there is a greater statistical chance of inter-meshing of hydration products with fine and coarse aggregate surfaces to produce a system of higher integrity on the micro scale and hence higher strength.

NORMAL WATER-REDUCING CONCRETE ADMIXTURES

This type of admixture is characterised by its ability to impart workability or permit water reductions without interfering significantly with the subsequent hydration of the cement.

Chemical type

These admixtures are usually based on either the calcium or sodium salt of lignosulphonic acid which is derived originally from the wood pulping industry. Modern admixtures based on this material are formulated from purified products from which the bulk of the sugars and other interfering impurities have been removed to low and consistent levels. The chemical formula of lignosulphonate is very complex, but can be said to be a polymeric sulphonate containing sufficient groups in the polymer backbone to allow adsorption on to cement particles without any significant interference with the hydration of those particles.

In addition to lignosulphonates there is another group of materials called polycarboxylic acids which find some application in this field and although limited in use at the present time do seem to offer for the future a means of providing "tailor made" synthetic admixtures.

The effect of calcium lignosulphonates in particular has been studied on the hydration products of Portland Cement (1) and it was found that although there is some evidence that the admixture may perhaps alter the rate of reaction, the products of hydration are unaltered.

Effect on the properties of plastic concrete

Set

By the very nature of their categorisation, admixtures of this type do not affect the set of ordinary Portland cement to a degree significant enough to make any changes in concrete placing practice necessary. At exceptionally high dosages there are indications (2) that the lignosulphonate material can cause retardation but these are at dosages many times that recommended by the manufacturers.

Air content

There is some indication that admixtures of this type may cause some slight increase in air content when used as a workability aid or as a water-reducing aid (2). However, this is usually only from 0.2 to 0.5% additional air entrainment and is more than compensated for by the additional strength possibilities. Also, materials are occasionally added to prevent this additional air entrainment.

Workability

Increased workability by the addition of a normal water-reducing admixture depends on a number of factors including the following:

(i) the dosage level of admixture
(ii) cement content
(iii) cement type
(iv) aggregate type.

Considerable data is available (3); generally it can be said that using a reputable admixture at the manufacturer's normal recommended dose and also at twice the recommended dose, one could apply the following workability data;

Table 2/2 Workability data for normal water-reducing admixtures.

	Control mix	Normal addition	Twice normal addition
Mix proportions:			
Cement (O.P.C.)) By	1.00	1.00	1.00
Sand (Zone 2)) weight	1.8	1.8	1.8
Gravel (20 mm - 5 mm))	4.05	4.05	4.05
Water cement ratio	0.55	0.55	0.55
Admixture	0	Normal dosage	2 x Normal dosage
Workability:			
Slump (mm)	76.00	133.00	178.00
Compacting factor	0.93	0.96	0.98
VeBe (secs)	1.8	1.0	1.0

Effect on the properties of hardened concrete

Strength development

The addition of a normal dosage of an admixture of this type can, as explained earlier, be utilised to produce a concrete having unchanged strength development yet increased workability or, alternatively, to allow water reductions to give the same workability with higher strengths. This latter application in turn can be utilised to allow cement reductions so that concretes can be more economically produced whilst retaining their strength and workability characteristics. Table 2/3 summarises the three possibilities in comparison with a control mix.

It will be noted that the strength development of mixes 2 and 3 is greater than that for the control, and in the case of mix 4 similar compressive strength results are obtained with a significant reduction in the proportion of cement used (approximately 30 kg/m^3 concrete).

Table 2/3 Examples of various concrete mixes with normal water-reducing admixtures.

Mix number			1	2	3	4
Purpose of use			Control mix	Increased workability	Increased strength	Cement economy
Mix proportions:						
Cement (O.P.C.)	)		1.0	1.0	1.0	1.0
Sand (Zone 2)	)	By weight	1.8	1.8	1.8	1.93
Gravel (20 mm - 5 mm)	)		4.05	4.05	4.05	4.55
Water	)		0.55	0.55	0.50	0.55
Admixture	)		0	Normal dosage	Normal dosage	Normal dosage
Properties:						
Density (kg/m^3)			2480	2480	2495	2480
Compressive strength (N/mm^2) at:		7 days	24.0	28.5	30.0	23.5
		28 days	30.5	33.0	38.5	32.5

Durability

The durability of concrete is its ability to maintain its aesthetic and constructional properties over a prolonged period of time. Any loss of durability can be caused by chemical attack, or by the action of freezing and thawing or wetting and drying etc. The most important factor influencing durability is said to be the surface permeability of the system (4). A considerable amount of work has been reported on the effect of admixtures on the surface permeability of concrete and it is quite clear that the straight addition of admixtures of this type does not cause any increase in permeability and, indeed, where the admixture is used to reduce the water cement ratio then permeability is considerably reduced. Many examples of more direct measurement of durability by freeze/thaw testing can be quoted and Table 2/4 illustrates one such example relevant to this type of admixture (3).

Table 2/4 Freeze/thaw durability and normal water-reducing admixtures.

Mix details: Cement content = 300 kg/m^3 approx.
Natural sand and crushed limestone. Slump = 100 mm approx.

		Freeze/thaw durability Reduction in dynamic modulus of elasticity (%) at				
		11	35	47	59	71 cycles
Admixture	*W/C Ratio*					
None	0.68	74	–	–	–	–
Normal dosage	0.55	19	33	37	39	43

Shrinkage

Admixtures of this type when used as workability aids or water reducers do not adversely affect the shrinkage, either due to the volume effect of hydration products under saturated conditions or due to moisture loss under non-saturated conditions, in comparison with mixes of similar workability and strength characteristics produced at higher cement contents.

Creep

Creep is an essential design factor and since it has been stated that it is associated with water movement at a molecular level then the surface active nature of admixtures has caused some concern as to its effect on this property. However, work has been carried out both in research establishments and in laboratories associated with major contractors to show that materials of this type have no deleterious effect on the creep of concrete.

Applications

The following list summarises the main applications for admixtures of this type:

a) the effective plasticising action of normal water-reducing admixtures increases the workability of most types of concrete mixes without reducing the compressive strength. This is particularly useful when concrete pours are restricted due to either congested reinforcement or thin sections.

b) when harsh mixes are experienced such as those produced with crushed aggregates then considerable improvements in the plastic properties of the concrete can be obtained.

c) when used for its water-reducing effect higher strengths at an early and subsequent age can be obtained. This may be applied where required strengths are difficult to obtain within specified maximum cement contents and where early lifting strengths are required.

d) by modifying the concrete mix design in conjunction with the addition of an admixture of this type then cement economies of about 10% can be obtained. As a rough guide it can be generally stated that for each 1p spent on an admixture then approximately 4-5p worth of cement can be saved.

ACCELERATING/WATER-REDUCING ADMIXTURES

Chemical type

These materials are generally based on calcium chloride which is included in the formulation to increase the rate of hardening at an early age. The other main ingredient is normally a lignosulphonate although hydroxycarboxylic acids in lower proportions can be utilised and tend to produce a solution less susceptible to sedimentation.

Effect on the properties of plastic concrete

Set

It will be seen from the British Standard requirement table quoted earlier that the stiffening time to reach 3.5 N/mm² of mortar sieved from the concrete should be at least 1 hr less than the control mix.

Air content

There is normally no increase in the air content of mixes containing admixtures of this type.

Workability

An accelerating water-reducing admixture can be used to produce a concrete possessing greater workability to allow easier placing in congested areas and yet provide high early strength. As an example, at a normal dosage of such a material then 40 mm slump would be expected to increase to approximately 70 or 80 mm. This is illustrated in greater detail in the section on Strength Development.

Effect on the properties of hardened concrete

Strength development

Development of high strength in the initial stages is the main application for admixtures of this type. Although the materials can be used as a straight addition to obtain higher workability with some increase in initial strength, it is more usual to combine some slight water reduction with the acceleration effect of the admixture to achieve very significant increases in strength. Table 2/5 illustrates the use of a straight addition of the material to produce workability with high early strength and, for general interest, a comparison is made not only with the control mix but also with a lignosulphonate plasticiser of the normal water-reducing type. Table 2/6 illustrates the use of the water-reducing capability of an accelerating/water-reducing admixture and again is shown in comparison not only with a control mix but also with calcium chloride solution itself.

Durability

The durability of concrete containing admixtures of this type can be discussed with regard to two separate and distinct areas:

1. Corrosion of reinforcement

Admixtures containing calcium chloride have been associated with the failures due to corrosion of reinforcements. In view of this CP 110, the code of practice for the structural use of concrete makes the following recommendation:– Due to the difficulties of maintaining a controlled dosage it is strongly recommended that calcium chloride should not be added to concrete containing embedded metal since its presence may lead to the accelerated corrosion of reinforcement.

It should, perhaps, be emphasised that calcium chloride accelerators are still permitted for the manufacture of components which do not contain embedded metal.

2. The permeability and/or the effect of the freeze/thaw cycling of the concrete

This aspect is not so well documented for accelerating/water-reducing admixtures as it is for normal and retarding/water-reducing admixtures. There is some evidence (6) that the effective pore diameter distribution is changed by the addition of calcium chloride and that there is some change in the hydration species of tricalcium silicates. The general picture is that as far as permeability is concerned it is reduced by calcium chloride and the net effect therefore of water reduction by the admixture and permeability reduction by the calcium chloride would be to produce a more impermeable concrete.

Shrinkage

When an accelerating/water-reducing admixture is used to produce high early strength with the simultaneous reduction of the water/cement ratio, in general it can be said that the shrinkage due both to hydrate volume changes and to moisture removal will be slightly higher than for a control concrete. Where a straight addition is made to provide a more workable yet quicker hardening concrete, then, there will be a greater increase in the shrinkages.

Table 2/5 Achievement of increased workability from accelerating water-reducing admixtures.

Admixture type	None	Normal water reducing	Accelerating water reducing
Mix proportions:			
Cement (RHPC)	1	1	1
Sand (Zone 2)	1.24	1.24	1.24
Gravel (20 - 10 mm)	3.20	3.20	3.20
Water	0.48	0.48	0.48
Admixture	—	Normal dosage	Normal dosage
Properties:			
Slump (mm)	38	64	69
28 day density (kg/m^3)	2480	2480	2480
Average compressive strength (N/mm^2) at			
7 days	33.5	32.5	40.5
28 days	49.0	52.0	56.5

Table 2/6 Achievement of maximum strength from accelerating water-reducing admixtures.

Admixture type	None	Calcium chloride solution	Accelerating water reducing admixture
Mix proportions:			
Cement (OPC)	1	1	1
Sand (Zone 2)	2.2	2.2	2.2
Gravel (20 - 10 mm)	4.5	4.5	4.5
Water	0.58	0.56	0.52
Admixture	—	Normal dosage (1.5% by weight of cement)	Normal dosage
Properties:			
Slump (mm)	13	13	13
28 day density (kg/m^3)	2450	2465	2450
Average compressive strength (N/mm^2) at			
7 days	27.1	29.0	32.5
28 days	42.0	45.5	53.2

Creep

There is little published data on the effect of admixtures of this type on the phenomenon of creep although what is available indicates that this type of admixture should not be used in creep-sensitive areas without fresh evaluations.

Applications

As outlined earlier, products of this type can impart workability whilst giving high early strength or alternatively even higher early strengths at similar workability to a control. In view of this, materials of this class will find application as follows:

a) to impart higher early and ultimate strengths to concrete, enabling forms to be stripped sooner in site-cast concrete and in the precast field for blocks and other items to achieve lifting strengths more rapidly. This effect is particularly useful at low ambient temperatures.

b) effective plasticising actions of this type of product will give increased workability to many concrete mixes to enable easier placing yet to raise early compressive strength. Water reductions can be made without adversely affecting the workability with consequent increases in strength, impermeability and durability.

RETARDING/WATER-REDUCING ADMIXTURES

This type within the water-reducing group retards the set and initial hardening of concrete whilst allowing increased workability or water reductions.

Chemical type

These products are based normally on hydroxycarboxylic acids which contain a considerable number of hydroxyl groups which are absorbed on to the surface of the cement, thereby retarding initial hydration. They also possess carboxyl groups which are negatively charged and cause the deflocculation of the cement particles as discussed earlier. In addition, lignosulphonates containing retarding ingredients are utilised, either fortuitously by the use of products of lesser purity (wood sugars) or intentionally by the addition of phosphate or sugar-based retarding species.

Effect on the properties of plastic concrete

Set

The ability to retard the set of concrete without adversely affecting the subsequent strength development is a major virtue of this type of material. The British Standard Specification states that the retarding influence at normal dosage should be at least 1 hr longer than for the control mix; this of course does depend on the mix ingredients, cement type, and dosage level. As a rough guide Table 2/7 indicates the extension of vibration time in relation to a control mix for four commercially available admixtures. The mixes were taken to constant workability:

Table 2/7 Vibration times using retarding water-reducing admixtures.

Admixture	Dosage	Vibration Limit*
–	–	100 (4 hrs)
1	Normal	150
2	Normal	200
	¾ x normal	165
3	Normal	160
	¾ x normal	170
4	Normal	140

*% of time without retarder at 60°F

In addition, work has been carried out (8) on the effect of the mixing sequence on the setting time when using retarding/water-reducing admixtures and it is important that a consistent procedure be carried out.

Air content

The hydroxycarboxylic acid type of retarding water-reducing admixture normally produces concrete having a slightly lower air content than that of a control mix. In the case of a lignosulphonate based material, then the air content might be 0.2 to 0.3% higher unless materials of the tributyl phosphate type are added.

Workability

Materials of this class in some cases have a much higher dispersing effect and hence water-reducing capability than the normal water-reducing type of admixture. In the case of hydroxycarboxylic acid type this is a basic function of the molecule itself and in the case of the lignosulphonate it is probably due to the higher dosages or solids content normally included in the formulation. Table 2/8 indicates the level of workability increase that can be expected from hydroxycarboxylic type admixtures.

Table 2/8 Workability increase from hydroxycarboxylic type admixtures.

Mix proportions:

Cement (OPC)	1	1	1
Sand (Zone 2)	1.8	1.8	1.8
Gravel (20 mm - 5 mm)	4.05	4.05	4.05
Water	0.55	0.55	0.55
Admixture	–	Normal dosage	2 x normal dosage

Workability:

Slump (mm)	76	152	177
Compacting factor	0.93	0.97	0.98
VeBe	1.8	1.0	1.0

Effect on the properties of hardened concrete

Strength development

The development of strength after the initial retardation of set with admixtures of this type and normal recommended dosages follows closely that of a control with no admixture addition. However, when water reductions are made then significant increases in strength are obtained which, as explained earlier, can again be used to allow cement economies in a mix. Table 2/9 illustrates the strength development of three different mixes in comparison with a control mix with no admixture additions.

Durability

There is considerable reported data (3) on the effect of hydroxycarboxylic acids and related derivatives on the permeability and freeze/thaw durability of concrete containing them. Table 2/10 illustrates the relative durability at a three month period of a concrete produced from natural sand and gravel at a slump of 75 mm with and without an admixture of the retarding/water-reducing type.

Shrinkage

As mentioned earlier, the volume change in the hardened state is due to, and is influenced greatly by, the amount of water and cement in the mix. The cement factor in itself has little direct effect except that it in turn affects the required water content of the mix; as the water content increases so does the tendency for drying shrinkage. The characteristics of the cement gel are thus primary governing factors and any step which can be taken to cause reduction of both water and gel formation is in the right direction. Water-reducing admixtures of this type are certainly useful in this context.

Creep

Until recently there was little experimental evidence to illustrate the effect of hydroxycarboxylic acids on the creep of concrete containing it. Recent work has shown that there are no indications of any deleterious effect on creep under saturated conditions.

Table 2/9 Strength development with retarding water-reducing admixtures.

Mix number	1	2	3	4
Mix purpose	Control mix	Increased workability	Increased strength	Cement economies
Mix proportions:				
Cement (OPC)	1	1	1	1
Sand (Zone 2)	1.8	1.8	1.8	1.93
Gravel (20 - 5 mm)	4.05	4.05	4.05	4.55
Water	0.55	0.55	0.50	0.55
Admixture	—	Normal dosage	Normal dosage	Normal dosage
Properties:				
Density (kg/m^3)	2480	2480	2495	2480
Compressive strength (N/mm^2) at 7 days	24.0	30.5	31.5	25.5
28 days	31.0	37.5	40.5	34.0

Table 2/10 Durability increase.

Admixture	W/C ratio	*Relative durability
None	0.61	26
Hydroxycarboxylic acid	0.58	56

*No. of cycles to reduce dynamic modulus by 50%.

Applications

Water-reducing admixtures which possess set-retarding capabilities are useful for obtaining increases in workability in those situations where an extension of time is required to place the concrete. Thus with a long and difficult pour the concrete can be successfully retarded for a number of hours and yet made to flow easily through congested areas. Applications and advantages, therefore, may be listed as follows:

a) It has been found that high strength, or high cement content, mixes in particular can be made workable by this type of admixture without loss in density, durability or strength.

b) Where long hauls of ready-mixed concrete are required then premature setting can be usefully avoided by the use of this type of admixture.

- c) When concrete is being placed or transported under conditions of high ambient temperature then again premature setting is avoided.
- d) Use of retarding/water-reducing admixtures make it possible to avoid the production of "cold joints" and is an aid in the slip-forming of mass concrete.

CONCLUSIONS

This chapter has presented information relevant to the three categories of water-reducing admixtures for concrete. It has attempted to be informative in terms of chemistry, mode of action and the effect on concrete in the plastic and hardened states. It is hoped that the publication of information of this type will be a contributive factor in encouraging the useful application of these materials both on site and in the precast yard to facilitate methods of working and to effect economies in the industry.

REFERENCES

1. S.M. Khalil, "Influence of Ligno-based Admixture on the Hydration of Portland Cements", Cement and Concrete Research, Vol. 3, 1973, pp. 677-688.
2. "Admixtures for concrete; Acceleration, Retarders and Water reducers", lecture at C. & C.A. Training Centre, 1972.
3. Symposium on Effect of Water-Reducing Admixtures and Set-Retarding Admixtures on Properties of Concrete, A.S.T.M., October 14th, 1959.
4. British Ready Mixed Concrete Association Technical Report Ref. TR6, August, 1973.
5. A.M. Neville, "Properties of Concrete".
6. Cement & Concrete Research, Vol. 3, 1973, pp. 689-700.
7. Humes Technical Bulletin, No. 4, February 1966.
8. Nature, Vol. 199, 1963, p. 32.
9. Cement Admixtures Association Wall Chart.

3
Air Entraining Admixtures for Concrete

K Bennett

INTRODUCTION

This chapter deals solely with the use of air-entraining admixtures in concrete.

A normal air-entraining admixture can be defined as any substance which can be utilised to intentionally entrain a controlled amount of air, in the right form, into a concrete mix without significantly altering the setting or rate of hardening characteristics of the concrete. It is, of course, possible to have air-entraining admixtures which also modify setting and hardening properties of concrete but these will not be considered.

It is necessary to distinguish between deliberately entrained air and that trapped accidentally. Entrained air takes the form of discrete bubbles whose magnitude is of the order of 0.05 millimetres to 1.25 millimetres, whilst we are all familiar with the appearance of concrete which has an excess of entrapped air.

PURPOSE OF AIR ENTRAINMENT IN CONCRETE

The primary purpose for the use of an air-entraining agent may be summed up in two words, namely "improved durability". Durability in this context is the increased resistance of the hardened concrete to the action of frost and to the detrimental effects of de-icing salts. Research has shown that for the whole of a concrete section to be adequately protected the bubbles must not be more than 0.4 millimetres apart. This is equivalent to the matrix of the concrete containing about 13% air by volume, which will be composed of that deliberately entrained and the residue of the entrapped air. Thus considering concrete we have typical air contents of 7% using 10 millimetre aggregate and 5% using 20 millimetre aggregate. Before we consider the probable mechanism of this improvement to durability it is worth mentioning that air entrainment can also be used to overcome the tendency for some concretes to bleed. Additionally, because the air bubbles behave in some respects as a fine aggregate, air-entraining agents are sometimes used to overcome inherent harshness in a concrete mix and to minimise segregation.

To appreciate how air entrainment benefits the durability of concrete it is necessary to have some understanding of the nature of hardened cement paste.

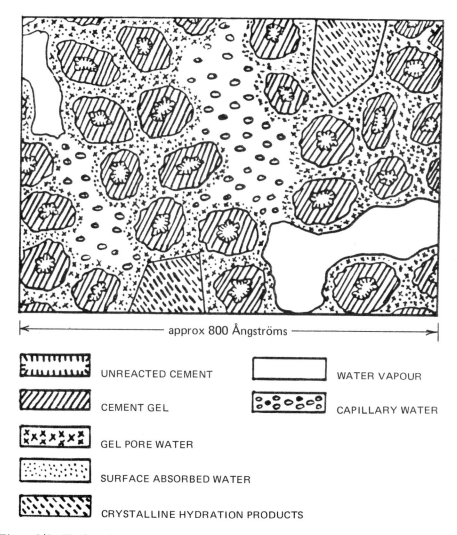

Figure 3/1 Hardened cement paste.

Figure 3/1 diagrammatically represents a section through hardened cement paste. In nearly all concrete mixes, for reasons of mobility and compaction, more water is used than is required to bring about complete hydration of the cement. In the hardened cement paste the excess water is variously absorbed on to the surface of the solid particles, located in cement gel pores or in capillaries distributed throughout the cement paste. It is not appropriate here to delve into the effects of these various phenomena as our interest is primarily concerned with the capillaries.

Problems of frost damage occur when these capillaries are filled with water and subjected to freezing temperatures. This can occur when a concrete is subjected to frost in its early life, before the capillary water is lost, or subsequently when concrete has been saturated by precipitation. The stresses, which have been calculated to be as much as 8 MN/m^2, set up by the expansion of liquid water to ice, exceed the tensile strength of the cement paste, with the inevitable result of failure.

The foregoing is a simplified explanation of the mechanism of frost damage but will suffice for our purposes.

The capillaries in the cement paste are random and inter-connected. It follows therefore that were these capillaries rendered discontinuous, the cement paste would be less liable to damage by freezing. Depending on the initial water/cement ratio adopted this will be achieved as the hydration process continues. For example, it has been shown that in uninterrupted hydration the capillaries are rendered discontinuous after three days at a water/cement ratio of 0.45, or six months at a water/cement ratio of 0.60. Above a water/cement ratio of 0.70 the capillaries can never achieve non-continuity. The implications are obvious; either concretes to resist frost damage are to have high cement contents or they are not to be subjected to adverse conditions for several months. The first is uneconomic except in special circumstances whilst the second is impracticable.

The use of air-entraining agents, however, offers an answer. It has been stated that the entrained air bubbles are approximately 0.05 millimetres to 1.25 millimetres diameter, some one thousand times larger than the capillaries. In air-entrained cement paste, therefore, the capillaries are interrupted by a relatively large void. Because of surface tension effects these voids cannot fill with water from the capillaries and therefore, under freezing conditions, behave as an 'expansion chamber' to accommodate the ice formed. When the ice melts surface tension effects draw the water back into the capillary so that the air bubble acts as a permanent safety valve, offering continued protection against frost damage.

Similar protection is offered when de-icing salts are used. These salts work by lowering the freezing point of water, the amount being proportional to the concentration of salt. For this to happen the salt must go into solution and this is accompanied by extraction of heat from the surroundings. One source of this heat is the water below the surface of the concrete. This transfer of heat in the solution process lowers the temperature of the sub-surface water to such an extent that it freezes. So although, for example, a road surface has been rendered ice-free for vehicles it has resulted in formation of ice in the bulk of the concrete. Once again the mechanism of the phenomenon has been simplified. It is known that the use of these salts can result in several cycles of freezing and thawing during one period of frost and therefore is potentially more disruptive to concrete than ice alone.

HISTORY OF AIR ENTRAINMENT

At this stage it is probably as well to have a brief look at the history of the use of air-entraining agents.

It is well established that the Romans used admixtures for their concrete. Such improbable substances as animal fat, milk and blood are recorded as being used to improve their pozzolanic concretes. One can speculate on the reasons for the use of these admixtures but it was probably to improve the workability of the concrete in its plastic state. However, blood for example is an air-entraining agent and, although the Romans were probably quite unaware of the fact, it is interesting to wonder if the well known durability of their structures can, at least in part, be explained by the improvements this effected in their concrete.

In more recent times, what may be described as the rediscovery of the benefits of air entrainment was quite accidental. To improve the clinker grinding process in the manufacture of cement various materials were investigated. One of the materials used as a grinding aid was beef tallow and it was subsequently observed that concrete made with this cement was far more resistant to frost attack than concrete made with cement in which no grinding aid had been adopted.

There is no question that initially air-entraining agents in common with most admixtures were received with some scepticism, possibly due to erratic results caused by insufficient proving tests and lack of understanding in their use. However we now have the situation where control of the products by the manufacturer and a better appreciation of the benefits to be obtained from their use have established admixtures within the field of concrete technology. In particular, air-entraining agents have been conclusively shown to improve the frost resistance of concrete. It will be demonstrated later that in certain circumstances air entrainment of concrete is often a mandatory requirement of some Specifications.

It is interesting to note that three grades of air-entraining cement for concrete are manufactured in the United States.

For reasons that will be dealt with in more detail later this approach to the production of air-entrained concrete is not favoured in this country, primarily because the amount of air entrained is so dependent on factors other than the amount of admixture used; that is, a fixed amount of admixture will not entrain a standard quantity of air for all types of concrete.

Current usage

The present Ministry of Transport Specification for Road and Bridge Works has a mandatory requirement that at least the top fifty millimetres of wearing course concrete paving is air-entrained. It can be safely assumed therefore that any major road of rigid construction will utilise some air-entrained concrete. Existing roads in the South of England using this construction material are sections of the M4 and M6 motorways, the A2 diversion in Kent, the works on the A12 in Essex, the M10 Hoddesdon by-pass, and the M27 motorway.

CHEMISTRY OF AIR-ENTRAINING ADMIXTURES

Basically all air-entraining admixtures come under the general description of surface-active agents (surfactants) which as the name implies involve a physico-chemical process occurring at the surface of constituent materials in a system. Surfactants are probably best known through their application by the detergents industry. In fact, in concrete technology, surfactants find application not only as air-entraining agents but also as plasticisers (wetting agents). It is not possible to draw a sharp distinction between the two but it will suffice to describe the former as those surfactants which will produce a stable foam with water and, in concrete, a stable dispersion of bubbles of the specified size and spacing. Both plasticisers and air-entraining agents will initially produce large numbers of small air bubbles, the difference being the stability of the foam thus formed.

The production of a foam depends upon reducing the surface energy of the water. It is not intended to go into the chemistry of surfactants in great detail, but it is assumed that it is known that a molecule of water is represented by the formula H_2O (Figure 3/2). The spatial arrangement is such that the water molecules are dipoles and have a strong affinity for each other, that is a great deal of energy is needed to pull them apart. This can be seen in the apparent gentle over-filling of a bowl with water before it finally cascades over the edge. This is a system having a high surface energy (surface tension).

THIS IS HOW THE ATTRACTION IS CAUSED —
YOU HAVE POSITIVE AND NEGATIVE IONS.

Figure 3/2 Representation of attraction between water molecules.

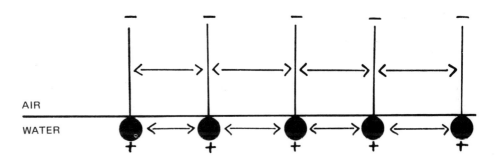

Figure 3/3 Surfactant abietic acid (vinsol resin).

Figure 3/4 Orientation of surfactant.

It has already been stated that the formation of a foam relies upon reducing this surface energy and surfactants achieve this in the following manner.

A surfactant is generally a long chain molecule (Figure 3/3), that is one having a 'skeleton' of many carbon atoms arranged linearly, one end of which is water repelling (hydrophobic), the other being water attracting (hydrophilic). When added to water the molecules tend to arrange themselves on the surface as shown in Figure 3/4).

41

Because of the orientation of the molecules of a surfactant, as shown in the figure, the attraction between them is less than that between water molecules, resulting in a reduction in the surface energy (that is, the energy needed to pull them apart). As has been previously stated the distinction between air-entraining agents and plasticisers is not clear cut. There are relatively few surfactants that will form stable foams in concrete but the following are the more commonly used:—

1. Animal and vegetable fats and oils.
2. Natural wood resins and their sodium salts, for example vinsol resin.
3. Alkali salts of sulphated and sulphonated organic compounds.

Manufacture and quality control

Whilst we are dealing with the chemical aspects of air-entraining agents it is appropriate here to briefly discuss their quality control during manufacture. As with all concrete admixtures uniform quality is a pre-requisite from the user's point of view. Most of the available air-entraining agents have their origins in resins and polymers which have certain inherent variations in their properties. It is the job of the manufacturer to assess the effect these inherent variations will have when the product is used in practice. This will be approached by carrying out exhaustive trials both on the admixture itself and on its performance in concrete with the final object of producing a material which will perform in a reproducible manner. Once these parameters are known and their effects appreciated, a Specification for the product can be written.

The Specification will include acceptance tests for the raw materials used in the manufacture of the admixture and for the uniformity of the marketed product. The latter tests would include, for example, moisture content, density and concentration of possible harmful constituents.

In addition it is perhaps worth mentioning that should problems be experienced with the products on site, most reputable manufacturers offer a comprehensive technical service which they are only too happy to put at the disposal of the user.

USE OF AIR-ENTRAINING AGENTS

In this section we shall be dealing with the practical considerations in the use of air-entraining agents, followed by some remarks on the design of air-entrained concrete mixes.

It has previously been stated that the distance between the air bubbles in the cement paste should not exceed 0.4 millimetres if complete protection against frost is to be achieved. It is essential that this air is uniformly distributed throughout the cement paste. It follows, therefore, that the air-entraining admixture must be an efficient foaming agent that will perform its function in a reasonable time in the mixer.

Normally the amount of air entrained in a given mix, for a set mixing time, is dependent primarily on the amount of air-entraining agent used, although there is a maximum dosage rate above which no increase in the volume of entrained air will be achieved. There are, however, several factors which will affect the amount of air entrainment obtained and we shall now consider these in general terms.

1. Richer concretes, that is those having higher cement content, require more air-entraining admixture to achieve the same air

content. Additionally more admixture is required with a more finely ground cement. It has also been shown that changes in the chemical nature of cement, which do not have significant effects on other properties, can alter the amount of air-entraining agent required.

2. The grading and nature of the fine aggregate used in the concrete will have a marked influence on the amount of air entrained. For example, it appears that an increase in the amount of material in the 300 - 600 micron size range increases the air content, whereas an increase of very fine material will decrease the amount of air entrained. An increase in the sand content as such brings about a small increase in the air content. Additionally the shape of the fine aggregate particles will play some part in the efficiency with which air is entrained into the mix. It must be appreciated that in the mixer, air entrainment is achieved purely by agitation. It follows that angular particles, which tend to cause more turbulence in the mixing process, will assist air entrainment.

3. For similar reasons more air is entrained if an angular coarse aggregate having a rough surface texture is used.

4. It is perhaps obvious that a foam cannot exist unless water is present. By an extension of this consideration it is clear that a concrete having a high water content will entrain more air than a similar mix at a lower water content.

5. The temperature of the concrete will affect the amount of air entrained. Air content varies inversely with the temperature.

6. It has previously been mentioned that entrainment of air relies upon agitation in the mixer. It follows that a more efficient mixer will entrain the desired air content in a shorter time. Consideration must be given, therefore, to the type of mixer available. It cannot be expected that a small free fall mixer will achieve the same effect in the same time as, for instance, a forced action pan mixer. It is also true that longer agitation times are necessary if small capacity mixers are used.

Whilst on the subject of mixers it is worth noting that severely extended mixing times will decrease the amount of air entrained into the concrete.

7. Air-entraining agents may be used in conjunction with other admixtures although their compatibility should be first ascertained. Once compatibility has been established it may be found that the amount of air-entraining agent used will need revision. For example, calcium chloride accelerators and pulverised fuel ash can both cause an increase in the amount of air-entraining admixture required to give a particular air content.

8. Finally, a word about transporting and placing air-entrained concrete. There appears to be a conflict of view on this matter, some authorities stating that entrained air can be lost during transportation (about 0.5 per cent) and whilst being compacted in place. If air is lost under these circumstances it is likely to be

in the form of the larger bubbles which are more similar to entrapped air and do not play a significant part in enhancing the durability of the concrete. Very extended vibration periods are likely to remove some of the entrained air but it is felt this will not occur during the period required to achieve a satisfactorily compacted concrete.

As far as pumping is concerned it has been found that air entrainment can be used with advantage where pipe lines are short and the pumping pressures will not be high. In longer pipe lines it must be remembered the pumping pressures are very high and may compress the entrained air to such a degree that it no longer functions in the intended manner, that is, to provide plasticity to the concrete.

DESIGN OF AIR-ENTRAINED CONCRETE

As with any concrete, correct mix design to achieve the requirements of a Specification is essential. A design must aim to produce an acceptable economic concrete that will not only meet the requirements in a hardened state but also one that can be handled and placed efficiently without excess bleeding or segregation.

It is well accepted that before any mix design is used for production concrete, it should be subjected to adequate full scale site trials. This procedure cannot be over-emphasised where air-entrained concrete is concerned. As we have seen, the mixer plays a large part in the amount of air entrainment obtained and to base site concrete on trials carried out in a small laboratory mixer, without including the site trial stage, is undesirable. Air-entrained concrete is normally specified in terms of compressive strength in the usual way, together with an acceptable air content range. For example, the Ministry of Transport Specification previously referred to limits the air content to 4½ ± 1½ per cent. Occasionally a limit is also placed on the water content of the mix.

Probably one of the best ways to approach an air-entrained concrete mix design is to initially arrive at a design for non-air-entrained concrete that will meet the strength, workability etc. requirements. Having designed or obtained a suitable mix the next state is to modify the proportions to allow for the entrained air.

It is well known that an increased void content in concrete causes a marked drop in its compressive strength. A general rule to follow for this is that a five per cent loss in strength will be incurred for every one per cent increase in air content. At first sight therefore it would appear that to entrain, say, five per cent air into concrete would have a catastrophic effect on its compressive strength properties. It will be remembered however that air-entraining agents are surfactants which also act as plasticisers (wetting agents). It has also been seen that the uniformly dispersed small air bubbles behave in some respects as fine aggregate. It follows therefore that a reduction in the water content of the concrete is possible due to the plasticising effect of the air-entraining agent and a further reduction can be made for a decrease in the fine aggregate content. The cumulative effect of this possible water reduction compensates to a large extent for the loss in strength incurred by the entrained air. It is likely, however, that an air-entrained concrete will have a slightly higher cement content than a non-air-entrained one, for a given strength. This aspect will be discussed in more detail when we look at the economics of air-entrained concrete.

Summarising, therefore, it will be seen that the non-air-entrained concrete mix design will need to be modified for water content, fine aggregate content and possible cement content.

Several authorities have proposed methods of calculating these modifications, amongst which is the one by Wright published in the Journal of the Institution of Civil Engineers (May 1953). A rule of thumb method was suggested by Shacklock in an article published in Surveyor (August 1960). He suggested a reduction in the sand content by 50 lb/yd^3 (30 kg/m^3) for every one per cent entrained air and then adding the appropriate quantity of air-entraining agent and water, by trial and error, to give a workability similar to that of non-air-entrained concrete. A method was also given for use where data for plain concrete was not available.

Whichever method is used it is essential that the proposed design is proved by site trials, with intermediate laboratory trials if considered desirable or necessary.

Having arrived at a suitable mix design and proved its performance by trials, consideration must be given to ensuring uniform production during the course of the contract. The amount of control necessary is likely to be only slightly greater than that which should be applied to any quality concrete production. The only extra operations which are involved are the accurate dispensing of the appropriate amount of admixture and monitoring of the concrete to ensure that the air content is within the specified limits. On the former point it is considered that the only satisfactory means is by the use of a suitable dispenser. Old milk bottles, tin cans or marked polythene buckets are unacceptable. Monitoring of the air content is readily carried out by use of an appropriate air meter, which is probably familiar to most concrete technicians.

BENEFITS OF AIR ENTRAINMENT

The primary benefit of air entrainment in concrete has been discussed at some length and it is appropriate to consider here the remaining attributes that can be achieved by the use of air-entraining admixtures.

These admixtures act as plasticisers in addition to their main purpose and will therefore improve the workability of concrete. They are not normally used as wetting agents, however, as a large range of alternative admixtures is available which will achieve this effect without air entrainment. The introduction of air into a concrete mix renders it more cohesive. This can be a great benefit when concrete needs to have a high workability but exhibits segregation and/or bleeding. This may be due to the nature of the constituent materials, and where economic alternatives are not available air entrainment is probably the only satisfactory remedy available. The effects of segregation can be quite serious both to the structural performance of the concrete and to its visual appearance, which may be important. Air-entraining agents can therefore be used additionally to enhance the visual quality of the concrete by improving its surface finish. This can be particularly valuable where fair faced concrete is specified.

Because the discrete air bubbles interrupt the continuity of the capillaries in the cement paste which, as has been seen, cannot fill with water, air-entraining agents reduce water absorption and permeability characteristics of the concrete.

Finally it is worth pointing out that the use of air-entraining agents will increase the yield of concrete. For a five per cent air content only ninety-five per cent of the concrete constituent materials normally used will be required to produce a given volume of concrete. This aspect will now be described in more detail, when further consideration is given to the cost of air-entrained concrete.

COST OF THE USE OF AIR-ENTRAINING AGENTS

The quantity of admixture needed to introduce the accepted amount of air into concrete, for complete protection, is small. For close batching control these admixtures are usually supplied as liquids which lend themselves to application by accurate dispensing equipment. The solids content of these air-entraining agent solutions varies between manufacturers but, as a guide, the amount of actual admixture needed may be as low as 0.04% by weight of the cement. It can be seen, therefore, why provision of a solution is essential for accurate batching.

It is difficult to quote an average price for admixtures but a typical figure for the cost of using the above amount of air-entraining agent would be about nine pence for concrete originally containing 300 kg of cement per cubic metre of concrete. It has been pointed out that the water content can be reduced to compensate for the increased workability and reduced sand content, which will to a large extent compensate for the loss in strength due to the entrained air. However, it was stated that a slight increase in cement content might be necessary. For the example quoted above an increase of about 15 kg of cement per cubic metre of concrete could be required. This represents an increased cement cost of about thirty pence. Against this can be set the saving due to the reduced sand content, in this case about twenty pence (4% reduction by weight). At this point a net cost increase of about nineteen pence per cubic metre of concrete is indicated. It must be remembered that an increase in yield due to the entrained air will be achieved. This increased yield will reduce still further the net cost increase. It will be seen therefore that what at first sight may appear to be a significant cost increase to achieve the benefits of air entrainment is in fact minimal.

FUTURE TRENDS OF AIR-ENTRAINED CONCRETE

In conclusion it is interesting to speculate upon the future use of air-entraining agents in concrete. Although the benefits to be obtained are now fairly widely appreciated, there is no doubt that 'pockets of resistance' still exist within the construction industry. Some of this can be attributed to the attitudes adopted by some Specifiers, either through lack of appreciation of the material or through adverse previous experience caused by its incorrect use or production. Certain manufacturers of concrete are reluctant to have another ingredient to batch and to monitor its effects, but the effort and capital equipment cost involved are small. These slight disadvantages are far out-weighed by the benefits to be gained.

It is quite likely that we shall see an expansion in the use of air-entraining agents for concrete roads etc. exposed to severe conditions and there is an arguable case for structural concrete similarly exposed. The general acceptance of air-entrained concrete, however, rests with an appreciation of its potential by all in the industry.

REFERENCES

1. J. Amer. Conc. Inst., 1971, No. 9, Proc. Vol. 68, pp. 646-676
2. "Concrete", 1968, Vol. 2, No. 1, pp. 39-44.
3. J. Amer. Conc. Inst., 1950, Vol. 22, No. 1, pp. 25-52.
4. J. Amer. Conc. Inst., 1954, Vol. 26, No. 2, pp. 113-146.
5. J. Amer. Conc. Inst., 1963, No. 11, Proc. Vol. 60, pp. 1484-1524.
6. "The Chemistry of Cement and Concrete", F.M. Lea, 1970, pp. 602-604, Edward Arnold (Publishers) Limited.
7. "Concrete Technology and Practice", W.H. Taylor, 1967, pp. 190, 250, 392, Angus and Robertson.
8. T.C. Powers, L.E. Copeland and H.M. Mann, P.C.A. Journal, May 1959, Vol. 1, No. 2, pp. 38-48.
9. Mobil Product EU 80Y, now Mobilcer HM.
10. F. Kocataskin and E.G. Swenson, ASTM Bulletin, April 1958, pp. 67-72.
11. M. Levitt, "The ISAT – A non-destructive Test for the Durability of Concrete", British Journal on Non-destructive Testing, 1970, Vol. 13, No. 4.
12. M. Vénuat, "Admixtures and the Treatment of Mortars and Concretes", published by the author in Paris.
13. "Freezing and Thawing of Concrete and the Use of Silicones", Highway Research Record, No. 18.
14. A.E. Desov, "The Effect of Silica-organic Compound GKZ-94 (Polyhydrosiloxane) on the Physico-technical Properties of Concrete and Mortars", Paper IV/7, International Symposium on Admixtures for Mortar and Concrete, Brussels, 1967.
15. A. Aignesberger and H.G. Rosenbauer, Tonindustried Zeitung, 1972, No. 96, pp. 29-34.

4

Pumping Aids for Concrete

A S Flateau and L H McCurrich

PURPOSE OF USE

The use of pumping aids during concrete pumping operations fulfils the following needs:

i) Increase in the range of mix designs which may be successfully pumped.

ii) Reduction in the risk of pipeline blockages.

iii) Improved flow of concrete through small bore pump lines, e.g. 75 mm diameter lines.

CONCRETE PUMPING PRACTICE

The volume of concrete placed by means of a concrete pump has increased over recent years. The advent of small bore mobile pumping units has made available to small or inaccessible sites the advantages of rapid concrete placement previously used only by large sites with large volumes of concrete to place. In addition, the view that concrete to be placed by means of a pump needs to be over-rich and of high workability is being modified as mix designs specially prepared for pumped concrete work have become available.

The economic use of a concrete pump has been questioned by engineers who have had experiences of pipeline blockages seriously interfering with the concreting process. The mix design for a particular job is critical for successful pumping operations but, having designed a suitable mix, one must also bear in mind that variations in the concreting materials themselves account for the major proportion of pipeline blockages. Cement and water do not significantly alter in their physical properties and each may be accurately batched into the mix. Aggregates, however, being less controllable materials, do vary especially in their gradings, shape and water content. It is in the broadening of the sources of aggregates which may be used in pumped concrete mix designs that admixtures have been most beneficial and, in addition, they reduce the effect of on-site variations.

Concrete is pumped through pipelines by either reciprocating piston pumps or peristaltic action pumps. In either case the concrete is pressurised and movement of concrete takes place under plug flow conditions, the central core of concrete sliding through an annular film of grout adjacent to the pipe wall.

The flow of concrete in pipelines has been investigated by a number of workers (1, 2, 3, 4) and it has been stated (5) that the lubricating grout film is typically about 3 mm thick (1/7 of max. aggregate size) for concrete flowing in a 100 mm internal diameter line at 30 m^3 per hour.

Pressures of 3.5 N/mm^2 may be developed in the pump and 2.1 N/mm^2 in the line. To maintain the grout lubrication under these conditions it is important that the water in the mix cannot readily be forced through the matrix of solid particles leaving behind a 'nest' of interlocked aggregate in contact with the pipe. This would result in excessive friction and pipe blockage.

The retention of water and the provision of an adequate grout layer is generally achieved by careful mix design, usually involving the addition of extra cement and sand. Admixtures may also be used to retain water and provide grout lubrication, and the type of admixture used will depend on the cement content of the concrete which may be considered in the following three broad classes:

1. Low cement content mixes with cement contents of up to 200 kg/m^3.

2. Median cement content mixes with cement contents of between 200 and 350 kg/m^3.

3. High cement content mixes with cement content in excess of 350 kg/m^3.

The problem most encountered in the pumping of low cement content mixes is a blockage brought about by a 'nest' of aggregate in the pipeline having been caused as described above. Concretes in the median range can be said to have grout consistencies which under controlled water contents will have optimum flow characteristics through the aggregate void channels, and thus this type of concrete is the one most associated with concrete pumping. High cement content mixes can product a blockage condition due to the high surface area of the mix constituents coupled with a low water/cement ratio producing a grout interface which exhibits high friction on the pipe.

Admixtures are currently being marketed to assist in the continuous pumping of all three types of concrete. In each case the admixture will modify the flow characteristics of the particular cement paste, helping to achieve and maintain optimum flow conditions. Research has shown that the complex regimes undergone by concrete flowing through a pipeline often make it necessary to resort to multiple types of admixtures and it is in this area that the compounding expertise of the admixture manufacturer is of value in producing materials tailor-made for the concrete engineer's use.

ADMIXTURE TYPES

Generally speaking, there are three types of pumping aids available on the market (2, 6):

(a) Thickening agent type

(b) Air-entraining plasticising type

(c) Non-air-entraining plasticising type

Water-thickening type

Materials of this type when added to water increase the viscosity of the water. Some may also tend to flocculate the cement particles thus providing a further thickening effect.

The chemicals used include polyethylene oxides, cellulose ethers and alginates. Very low concentration can produce significant increase in viscosity as is illustrated on the graph for one cellulose ether. (Figure 4/1).

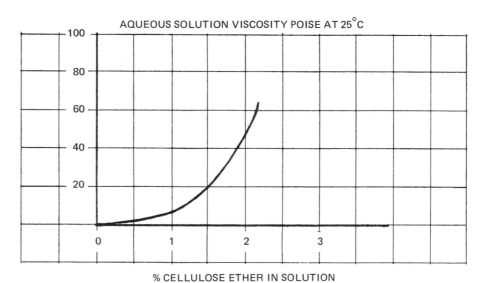

This curve is for one particular type of cellulose ether. Many grades are available covering a wide range of viscosity characteristics.

Figure 4/1 Viscosity graph.

Air-entraining plasticising type

These are made from natural or synthetic materials which have surface-active structures. Air entrainment is achieved by chemicals which lower the surface tension of the water enabling microscopic sized bubbles to be formed. The bubbles are stabilised within concrete by the chemicals forming physico-chemical linkages between bubbles and cement. Examples of materials used are neutralised wood resin and alkyl sulphonate. Air entrainment will itself plasticise the concrete but additional plasticising chemicals may also be included in the admixture manufacturer's formulation.

Non-air-entraining plasticising type

Chemicals used in admixtures of this type operate by providing better dispersion of cement particles in suspension. They are surface-active agents which are absorbed on to the cement particles, leaving a partly charged group on the surface which repels similar charges on the adjacent cement particles.

Materials used in this type of admixture include lignosulphonates, salts of hydroxy carboxylic acids, melamine formaldehyde condensates and sulphonated napthalenes.

USE OF PUMPING AIDS IN CONCRETE MIXES

As explained earlier, the performance of a mix in a pump will depend upon the cement content, aggregate/sand type, shape and grading, and water content. For pumping mix design, the simplest way of looking at these variables is by considering the void content between the aggregates and sand particles in relation to the volume of cement paste required to fill these voids (7, 8). A degree of overfill is generally required for workability and durability. For pumping, the degree of overfill is often further increased and some authorities recommend that the fines content (cement plus material smaller than 0.2 mm) should be at least 350 kg/m³ for 20 mm max. size aggregate and 400 kg/m³ for 10 mm max. size aggregate. It may not always be economic to meet these optimum requirements and pumping aids are especially useful in mixes of low cement content where the required overfill may not be achieved.

Methods of testing concrete for ease of pumping are being developed. The significance of the rate at which water can permeate through a mix has already been discussed. A direct method for measuring this is the pressure bleed test (4, 7) in which a small quantity of mix is placed in a pressure pot and the amount of bleed water which comes through a filter membrane at given time intervals is measured. This test can be used to investigate the performance of admixtures, especially those of the water-thickening type.

The types of pumping aid used will now be considered in relation to the broad classification of cement contents given earlier.

Low cement content mixes (up to 200 kg/m³)

Perhaps the most significant contribution that admixtures have made to the pumping of concrete is the ability to use concretes with cement contents in the region of 130 kg/m³. In such concretes, without the use of admixtures, the flow of grout through the void channels is extremely rapid and escape of grout occurs under pumping pressure along the concrete pipeline interface. When such grout loss occurs, an aggregate nest is formed creating a typical blockage.

The use of a thickening agent admixture reduces the problem of pressure segregation and typical results quoted by Kempster (2) are given in Figure 4/2. This illustrates how the range of pumpable concrete admixtures may be extended by the use of a pumping aid.

VOID CONTENT VOLUME

CEMENT CONTENT BY VOLUME	17% (1)	20% (2)	25% (3)	28% (4)	
Z 15%	PUMPED	PUMPED BLOCKED DURING EXPERIMENT	DID NOT* PUMP	DID NOT PUMP	Sand content was 35% of combined aggregate in all mixes
Y 20%	PUMPED	PUMPED	DID NOT* PUMP	DID NOT PUMP	Water content was adjusted to give 75 mm slump
X 25%	PUMPED	PUMPED	PUMPED	DID NOT* PUMP	* Mixes which became pumpable when cellulose ethers were added
W 30%	PUMPED DIFFICULT	PUMPED	PUMPED	DID NOT* PUMP	

Figure 4/2 Pumpable concretes.

Median cement content mixes (200-350 kg/m³)

Many concretes with the median range of cement contents will have satisfactory cement paste flow properties. It is possible, however, that consistent supplies of coarse aggregate and particularly sand may not be easily obtained or may vary in the course of a pumping operation. Such changes in the aggregates may adversely affect the balance between grout volume and aggregate void space, thereby reducing the stability of the concrete to the pumping conditions. This is particularly true of cement contents in the lower and higher echelons of the median range and in these regions difficulties may be encountered in achieving satisfactory flow. With these conditions in mind, admixtures have been designed to fulfil a multiple role. By using admixtures of the air-entraining plasticising type it is possible to release mixing water to the cement paste and simultaneously to control the consistency. Concretes in the lower echelon of cement content, within the median range, in which the aggregate void pattern would allow rapid grout flow and loss, are modified by containing the free water within the cement paste and not allowing the consistency of the cement paste to fall to the unstable limit.

With air entrainment the air content should not exceed 5%, because compression of this air in the pipeline will reduce the output of the pump, especially in very long lines.

In the richer mixes the release of mixing water to the cement paste allows more adequate flow and hence a reduction in the tendency for friction to develop in a grout film of too stiff a consistency.

An example of the increase in cohesion without excessive sand contents which can be achieved with a pumping aid of the air-entraining plasticising type is given in the following table.

Table 4/1 Air entraining plasticising type admixtures – mixes with acceptable cohesive properties for pumping: cement content 325 kg/m³.

Pumping aid dose per 50 kg cement	Sand Content %	W/C ratio	Slump mm	28 day strength N/mm²
None	44	0.61	50	43
0.14 litre	39	0.56	75	49

High cement content mixes (above 350 kg/m³)

The third category of concretes, namely those of high cement content, tend to have cement pastes of stiff consistency and do not have the required lubrication characteristics. The grout film formed does not easily allow the central plug of concrete to move down the pipeline and low pumping speeds result. In the extreme case, the total frictional force balances the pumping pressure and the flow of concrete is stopped. An example of this is given in Kempster's diagram (Figure 4/2). Admixtures used for improving the pumping characteristics of rich mixes often possess a plasticising action which will release extra mixing water to the cement grout. The high cement concentration alleviates the necessity to fix the extra water, due to the hydrophilic action of the cement, and the non-air-entraining plasticising type of admixtures are currently being used as pumping aids for this class of concrete.

PRACTICAL CONSIDERATIONS

The water-thickening type of pumping aid is supplied in powder or liquid form. No cheap reliable dispensers are yet available for dispersing very small quantities of powders into mixers, and the admixture suppliers' advice must be taken. This will involve either the use of small pre-packed quantities or the provision of a volume cup of the correct size for a given quantity of cement.

The addition of the air-entraining and non-air-entraining plasticisers is a relatively simple operation for, as liquids, they can be put through a number of dispensers which are commercially available.

FUTURE TRENDS AND ECONOMICS

Estimates with regard to the increasing use of pumped concrete vary but there is no doubt that in the long term a substantial quantity of the concrete produced in this country will be pumped. Added to this factor is the limited supply and increased consumption of higher quality aggregates which must result, in the long term, in lower quality materials having to be used. The latter materials will necessitate the use of admixtures to facilitate concrete pumping. The use of admixtures for pumping is not one that can easily be brought down to a pounds and pence situation. It is more to the point that, by using admixtures, a wider range of cement contents and aggregates can be used and the pipeline friction can be reduced.

CONCLUSIONS

At present the use of special pumping aids in concrete is limited. The admixture type used most widely in pumped concrete is the normal non-air-entraining plasticising type. The use of special admixtures of the thickening agent type or air-entraining plasticising type may well increase in the future if there is increased demand for pumping lightweight aggregate concrete and concrete with lower cement contents made with poor aggregate gradings.

REFERENCES

1. I.R. Weber, "The Transport of Concrete by Pipeline", Cement & Concrete Association, translation No. 129.
2. E. Kempster, Building Research Station Current Paper 29/69, August 1972.
3. R.E. Tobin, "Hydraulic Theory of Concrete Pumping", ACI Journal, August 1972.
4. F. Loadwick, "Some Factors Affecting the Flow of Concrete through Pipelines", Proc. 1st Int. Conference on the Hydraulic Transport of Solids in Pipes, BHRA Cranfield, September 1970.
5. J.D. Parkinson, "Workability and Flow Behaviour for Pumped Concrete", Concrete, December 1971.
6. "Placing Concrete by Pumping Methods", ACI Committee 304, Report No. 68-33, ACI Journal, May 1971.
7. E. Kempster, "Concrete for Pumping", Concrete, December 1971.
8. R.D. Browne and P.L. Domone, "The Long Term Performance of Concrete in the Marine Environment", Proc. of Conference on Offshore Structures, Inst. of Civil Engineers, October 1974.

5

Integral Waterproofers for Concrete

P C Hewlett, R W Edmeades and R L Holdsworth

It is claimed by some that concrete admixtures are not necessary and are certainly no substitute for sound concrete mix design. Others argue that addition of admixtures offsets errors at the practical stages of making concrete, so increasing the chance of producing concrete up to specification. These statements are contentious, but perhaps have some justification for that group of admixtures known as integral waterproofers.

The need for waterproofing admixtures

There is no doubt that good workmanship coupled with correct mix design can give rise to well-made concrete which has low permeability and porosity (typically K_w* = 10^{-10} - 10^{-12} cm/sec). If the concrete is also adequately cured it may be regarded as water-tight since passage of moisture and water through the concrete is less likely, due to discontinuity of capillaries and pores. However, we are all familiar with the fact that in practice concrete structures often allow the passage of water, not only through joints and discontinuities, but through the mass concrete itself.

A study of practical situations will show that water can penetrate concrete in two ways. Firstly, when there is a hydrostatic pressure on one side of a concrete mass, water and aggressive agents can travel through any channels which interconnect the two faces of the concrete. Secondly, water may be absorbed by capillary action and travel through the concrete to a face where it evaporates because the air in contact with that surface is unsaturated.

Admixtures that reduce permeability will be effective in the first situation. However, admixtures that impart water-repelling or damp-proofing properties may reduce the effect of the second mechanism whilst being less effective at preventing water passage under a positive hydrostatic head.† However there are instances where hydrophobing waterproofers prevent penetrating damp under modest wind pressures.

A common factor in all concretes allowing the passage of water and/or water vapour is the presence of inter-connected voids. Without such voids and their interconnections water or vapour transfer cannot take place. Let us consider for a moment how voidage in concrete can be altered.

One way in which the number of channels and capillaries can be reduced is by controlling the water/cement ratio. During the hydration process of

---

* K_w refers to the coefficient of hydraulic permeability.

† The capillary suction generated by dry or partially dry concrete in contact with water can be equivalent to a hydrostatic head of several metres (Ref. 11).

the cement some of the capillaries need only become partly blocked by hydration products ("Tobermorite" or cement gel — $3\,CaO.\,2SiO_2\,.\,3H_2O$). For ordinary Portland cement concrete if the water/cement ratio is in excess of 0.70 there will not be sufficient gel formed to block the capillaries resulting in interconnections. The cement gel itself has very low hydraulic permeability ($K_w = 7 \times 10^{-14}$ cm/sec). ((8) and Figure 5/1).

The data given above is for cement pastes and the addition of aggregate changes the values slightly. In many practical situations these requirements are unfortunately seldom attained and the use of an admixture may help to restore the balance. This is probably most pertinent in leaner mixes where the cement content is less than 250 kg/m³. However, malpractice continues and doubt exists about the need for and specification of waterproofing admixtures.

Perhaps the reasons for this unsatisfactory state are,

(a) the difficulty of reliably demonstrating the imparted property of waterproofness, and

(b) the multifarious range of materials differing chemically and in mode of action and yet seemingly resulting in virtually the same property.

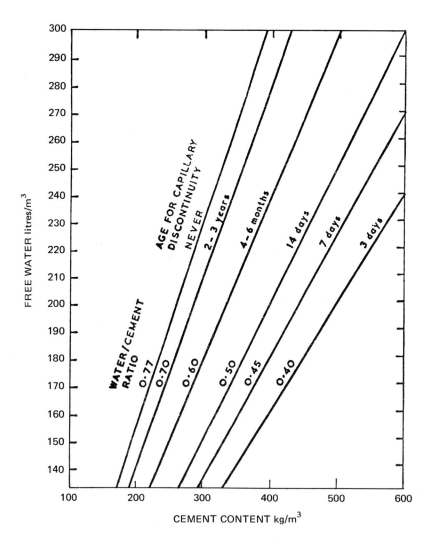

Figure 5/1 **Capillary continuity or discontinuity in cement pastes.**

In short, there is a general disbelief in the efficacy of these materials. It is interesting to note that integral waterproofers form the largest category of products in both the CAA and BRMCA Admixture Charts.

The history of waterproofing admixtures is sparse and the use of chemicals having waterproofing properties seems to have preceded any detailed understanding of their mode of action. Vénuat has reviewed the history of admixtures (12). Waterproofers probably developed in an attempt to upgrade lean concretes at a time when cement was a very expensive material. Also, particularly in the U.S.A., very wet mixes were at one time popular and their use was quite likely to result in porous and permeable concrete.

Definition of an integral waterproofer

There have been many attempts at listing and updating lists of waterproofing admixtures and several references are given (1-7). It is important for the purpose of this chapter to define what is meant by an integral waterproofer. We consider an integral waterproofer to be a material (powder, liquid or suspension) that when intimately mixed with the fresh concrete results in,

(a) reduction in the hydraulic permeability of the cured concrete mass and/or

(b) a water-repelling or hydrophobic property being imparted to the set concrete.

This definition excludes accelerating admixtures such as calcium chloride since such materials only alter the rate at which some initial permeability is reached and not necessarily its ultimate value. Additionally, plasticising or water-reducing admixtures should also be excluded since, although there is adequate evidence to show that total porosity may be reduced slightly by their use, their effect on permeability has not been established beyond doubt. However, since several proprietary waterproofers belong to this Group, they are given further mention below. Finally, waterproof cements are omitted since these are made by blending the waterproofing chemical, such as metal stearate, non-saponifiable oils and treated gypsum, with the dry cement powder. By definition such materials are additives and not admixtures. Likewise any *surface* treatment that imparts water-repellency or waterproofness is not dealt with since these materials are applied after the concrete has set.

CLASSIFICATION OF WATERPROOFING ADMIXTURES

Accepting that there is a fundamental need for a material that will upgrade the inherent water stopping properties of concrete, what materials are available? How do they function? In what way are they used?

Firstly, we may classify waterproofing admixtures into three main groups,

(i) Permeability reducers

(ii) Water repellents or hydrophobers

(iii) Miscellaneous

Members of each group may be reactive or non-reactive. By reactive we mean that when the admixture contacts wet cement chemical interaction occurs resulting in a new product which imparts the waterproofing property. Non-reactive materials will include the group known as "inert fillers".

Integral waterproofers range from thick, clear or cloudy liquids, to pastes or dry powders. They may be added to the gauging water or to the concrete mix but, as previously stated, materials that have been pre-mixed with the dry cement are excluded.

Group (i) admixtures will be both porosity reducing and pore-filling materials. This is not usually so for Group (ii) materials. Each Group is dealt with in turn below.

Permeability reducers

We can sub-classify materials within this Group, which reduce the hydraulic permeability of concrete, into

(a) very fine particulate materials

(b) workability and air-entraining admixtures

(c) accelerators.

Below, we deal with each sub-group in turn.

Very fine particulate materials

Ground sand, whiting, bentonite, PFA, diatomaceous earth, limestone, slag and pumice, colloidal silica and fluorosilicates comprise, in part, this group. It is preferable for the finely divided material to have some degree of pozzolanic reactivity, so densifying the cement gel matrix by the replacement of coarse calcium hydroxide crystals with finer gel-like hydrated calcium silicate products.

Particulate materials are of real benefit if the concrete mix is low in cement or deficient in fines. However, in cement rich mixes the effect could be the reverse since the addition of fine particles could increase the water requirement, leading to a less dense and lower strength concrete. It could be argued that so long as heat evolution is not a problem and shrinkage cracking can be avoided the addition of extra cement to lean mixes (refer CP 2007) is the best means of waterproofing.

Workability and air-entraining admixtures

Incorporating a workability agent or plasticiser reduces the chance of large voids and the lower water requirement offsets bleed. Salts of lignosulphonic acids are commonly used, often in combination with a particulate mineral filler.

The lignosulphonate is absorbed on to the C_3A early hydration products, probably giving better dispersion of the cement particles. At the same time the surface tension of the aqueous phase is reduced, perhaps aiding better wetting and reaction. Incorporation of lignosulphonates may give rise to a finer pore system which, whilst causing high capillary absorption, increases resistance both to actual passage of water and, hence, to permeability.

Air-entraining agents act in a similar manner to lignosulphonates by imparting improved workability to the mix and thus allowing less water to be used. The cellular structure does not give rise to inter-connected voids which would otherwise increase the permeability.

Common materials in this group are neutralised wood resins, detergents and sulphonated carbohydrates, anionic surfactants in general.

Accelerators

The function of accelerators as permeability reducers is doubtful. The use of calcium chloride, for instance, may improve early permeability and porosity figures simply by advancing the overall hydration reaction of the C_3A and C_3S phases but the ultimate permeability remains, as far as the authors are aware, unchanged.

Water repellents or hydrophobers

Materials in this Group reduce the passage of water through dry concrete which would normally occur as a result of capillary action and not as a result of an external pressure of water. Moisture movement of this type may result in staining efflorescence and general unwanted dampness.

In principle it is thought that all these materials impart a water repellent property to the concrete surface as well as lining and, in some cases, blocking the pores. The detailed mechanism is obscure but it has been suggested (2) that the water repellent action is associated with an electrostatic charge imparted to the walls of the capillaries. Materials comprising this group are,

(a) Soaps

(b) Butyl stearate

(c) Selected petroleum products.

Soaps

These are usually metal salts and, more recently introduced, the sulphonium salts of fatty acids, for instance calcium, sodium or ammonium stearate or oleate as well as stearic acid dispersions. The "soluble" soaps are thought to react with the calcium ions in the aqueous phase of the concrete and so precipitate out as the insoluble calcium salts. It is this precipitate that imparts the hydrophobic coating to the capillary surfaces as well as blocking some pores in the fresh concrete. The pore system that develops during the later stages of hydration (greater than 24 hours) is not affected by this precipitate, hence the saturated permeability is not reduced.

Many of these materials entrain air, due presumably to their surfactant properties, and dosages exceeding 0.2% m/m of cement are not recommended without the likelihood of significant strength loss. On the other hand, improved workability may result from using these soaps, which offsets the formation of cavities and large voids.

The optimum dosage should be established taking regard of the mix design, admixture chosen and final properties required.

Butyl stearate

The hydrophobic action of the butyl stearate is similar to that of the soaps, in that the eventual compound resulting in water repellency is calcium stearate.

However, butyl stearate hydrolyses only slowly in the alkaline phase of concrete and slowly produces the calcium stearate. As a result very much less air is entrained and strength reductions are not so serious. This allows higher concentrations to be used, which invariably gives improved damp-proofness. Slow reaction of the butyl stearate also allows better distribution of the admixture throughout the cement gel.

Butyl stearate is usually added as an emulsion at a dosage that gives 1% w/w butyl stearate to cement.

Selected petroleum products

Mineral oils, waxes, cut-back and emulsified asphalts comprise this group. In the case of the asphalt emulsions the dispersion is broken by the drying out of the concrete, resulting in hydrophobing and some pore blocking. The other compounds are generally regarded as "inert" hydrophobers acting mainly in a physical way without obvious reaction with the cementitious components.

Some strength reduction may result due to entrained air, particularly if emulsifying agents have been used to disperse the hydrocarbon. Certain wax emulsions may improve the damp-proofness of steam cured concrete (9) by melting "in situ" and so blocking the pores.

Miscellaneous

Products in this group are considered to have some waterproofing function but the chemistry is obscure and practicability perhaps doubtful.

Examples of such materials are petroleum jelly, napthalene, wax, certain celluloses and diluted coal tars and, finally, sodium silicate and aluminium powder.

Proprietary products comprising any one of several materials from this group have to be judged on their *proven* merits. Justification may often appear obscure and no strict recommendation can be given.

Some proprietary waterproofers are mixtures of several materials derived from two or more of these groups and may be regarded as multi-functional. Such admixtures seek to reduce permeability and impart "damp-proofness" to concrete.

Silicones, although widely used for the surface treatment of concrete, have not found much use as integral waterproofers. The cheaper silicones tend to cause excessive retardation when used in sufficient quantity to affect the properties of hardened concrete.

Methyl siliconates and polyhydrosiloxanes have been evaluated for improving the frost-resistance of concrete (13, 14) but the permeability was not found to be reduced. Research on these materials continues.

TESTING METHODS

Establishing the efficacy of integral waterproofers has always been a problem. The very real difficulty of measuring the hydraulic permeability of the concrete to obtain steady and consistent results is well-known (6, 10). It is probably as a result of this difficulty that practical test methods have evolved on a rather ad hoc basis and as a general rule waterproofing test data is rather poor. Most admixture manufacturers' data relate to a simple absorption or immersion test (refer British Standard 556 part 2:1972) or a partial hydraulic permeability test. In many cases this latter test relies on making a "permeable" concrete to begin with which serves as the control. Just how permeable the concrete can be and still benefit from the incorporation of an admixture is a matter for argument. These ad hoc tests are mentioned below.

It is recommended that more attention should be paid to the adoption of some of the more ingenious laboratory test methods reviewed by Kocataskin & Swenson (10) for measuring the effects of the integral waterproofers. Those tests are more meaningful in the absolute sense and allow performance comparisons to be made. They cover both the "damp-proofing" and "permeability reducing" categories.

The various available methods are dealt with below.

A marked difference between all the methods is the initial state of the sample before the testing, quite apart from variations in the test methods themselves. We have attempted in Figures 5/2 and 5/3 to show what the test methods actually measure and the initial condition of the concrete.

Test methods for damp/waterproofers

Method I

This method is that detailed in British Standard 556: part 2: 1972 or a modification of it. Essentially the concrete to be tested is oven-dried and

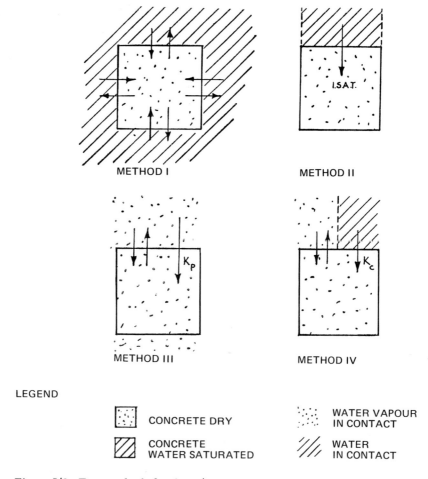

Figure 5/2 Test methods for damp/vapour proofers

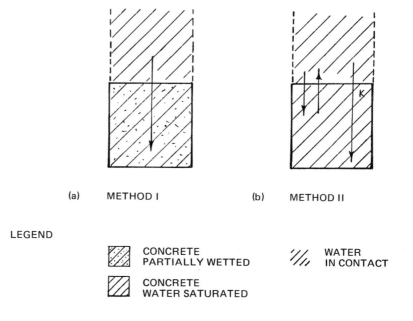

Figure 5/3 Test methods for waterproofers.

immersed in the dried out state in water for fixed periods of time ranging from a few minutes to many hours. Water absorption is measured as the weight change at the fixed immersion times.

$$\text{Absorption (\% of dry weight)} = \frac{(\text{wet weight - dry weight}) \times 100}{\text{dry weight}}$$

This test has some merit, being very simple and giving results that may be compared one with the other on relative performance. However, the information cannot be related to any absolute property.

Method II   This method is widely known as the Initial Surface Absorption Test or ISAT and is contained in British Standard 1881 part 5: 1970 "Methods for Testing Concrete". The test measures the rate of flow of water into concrete per unit area after a stated interval from the beginning of the test. A small constant head of water is used and a steady temperature maintained. The concrete is again oven-dried before carrying out the test. The apparatus is shown in Figure 5/4.

Briefly, the cell or cap (of known area) is clamped or fastened to the surface of the concrete and filled with water from the reservoir. The reservoir supplies a constant head of water of about 20 centimetres. After entrapped

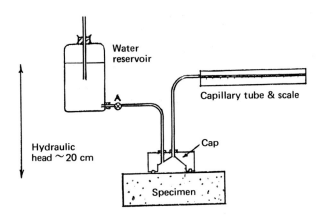

Figure 5/4 Initial Surface Absorption Test.

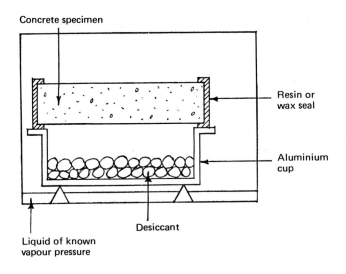

Figure 5/5 Test for K_p determination.

air has been bled off through the central connection the tubing is quickly connected to the calibrated capillary and this is flushed to expel all air.

At the start of taking a reading, tap (A) is closed and the movement of the meniscus in the capillary tube is then timed between two selected calibration points. Alternatively the distance moved through a given time is noted. Readings may be taken at time intervals ranging from a few minutes to several hours. ISAT values are quoted in units of ml/sq metre/second.

Method III (reference 10)

This method really measures the unsaturated steady state coefficient K_p where water vapour is the migrating fluid.

$$Q \text{ (cc/sec)} = K_p \frac{dp}{dx} \quad \text{(Per unit area)}$$

Where dp/dx refers to the vapour pressure potential or driving force, K_p is measured using variations of the dry cup* test. The technique is shown diagrammatically in Figure 5/5.

Tablet samples of concrete, mortar or cement paste are sealed at the edge with wax or synthetic resin. The concrete is placed in a chamber containing water vapour at known vapour pressure, the sample is allowed to equilibrate, after which it is attached to an aluminium cup containing desiccant such as anhydrous calcium chloride.

It is important to maintain constant temperature. Moisture diffuses through the pores of the concrete and is taken up by the desiccant. Weighing the cup and specimen at regular intervals shows when the "steady state" has been reached. Also by storing specimens in conditions of different relative humidity, values of Q against vapour pressure P may be obtained. Plotting values of Q x X (X = specimen thickness) against P we obtain a smooth curve represented by the function

$$\int_0^P K_p \, dp$$

By graphical differentiation the unknown coefficient Kp may be obtained. Plotting Kp against vapour pressure often gives a very good semi-quantitative indication of performance of one admixture against another. Unfortunately the method is restricted to small specimens.

A rather special case of water transport through a specimen is when we have liquid on one side and vapour transport through the sample and out the other side. Here we have a condition of saturated and unsaturated permeabilities. It is possible by using a dry cup upside down and having water on the back surface to obtain some measure of liquid/vapour transport through the specimen. What data exists implies that transport occurs only by vapour and that most admixtures studied appear to have no effect on the saturated permeability!

The liquid/vapour transport in concrete is probably the most pertinent one in practice and yet seems to be the least studied.

---

* Similar methods are used for determining the vapour permeability of paint film or membranes.

Method IV

This method measures the "capillarity coefficient" K_c which in turn is a function of porosity (E), permeability (K) and "suction" of the material (F). These variables are also related to the moisture content at any point X from the surface by the equation,

$$K_c t = (2_e\ EKF)t = \theta^2$$

where e = density of water
and t = time interval

K_c is *not* a constant and depends on the initial moisture condition of the concrete. Samples are equilibrated at a chosen relative humidity then weighed and placed in contact with water on one surface only. The amount of absorption is measured by taking weights at various time intervals.

Plotting the weight gain against time gives straight lines that may be represented by the equation,

$$\theta^2 = K_c t + a$$

where 'a' represents the amount of water required to initially wet the surface. Slopes of the curves give K_c (g²/cm⁴ × minutes). The capillarity coefficient values with the relative humidity give a Sigmoid pattern and again families of curves have to be compared in order to establish the general and particular effect of an admixture. It is imperative to test against controls.

Test methods for waterproofers/ permeability reducers

Here we are concerned with ways of measuring the effect of an admixture on altering the flow coefficient K_w or hydraulic permeability. This saturated steady flow coefficient is related to flow-rate, etc., by the Darcy formula

$$Q\ (cc/sec) = K_w \frac{\Delta p}{\Delta x}$$

where K_w has units of cm/sec

and $\dfrac{\Delta p}{\Delta x}$ is the applied pressure gradient

As with the other flow coefficients K_p and K_c, ad hoc testing methods have developed due in part to the difficulty in measuring K_w reliably enough to use it as a diagnostic of admixture performance. However, two techniques are described below.

Method I

Porous concrete* tablets (5″ x 2″) containing an admixture are cast, cured and at 28 days mounted in an enclosed cell, and water pressure applied (Figure 5/3(a)). The pressure is increased in stages and maintained for a short period (hours, minutes) at each pressure. The under surface of the tablet is observed for the appearance of moisture droplets. The pressure at which beads of moisture appear is noted and compared against a control sample cast from the same batch of concrete, but containing no admixture. At best this method gives qualitative comparisons and can only relate to short term behaviour. It is, unfortunately, quite widely used.

Method II

This consists of actually trying to measure K_w, and one variant of the apparatus is shown schematically in Figure 5/6. Concrete specimens are cast

---

* A typical porous mix contains ¾″ aggregate, sand and cement in the proportions 3.7:2.5:1, and has a W/C ratio of about 0.63 to give a 2″ slump. Air curing in a humidity cabinet is employed.

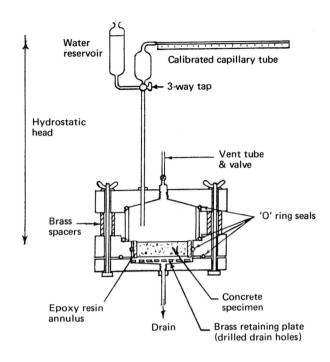

Figure 5/6 Saturated permeability test.

as a cylinder or core and then sliced in thicknesses of about 2.5 - 4 cm. The tablets are then set in an annulus of epoxy resin and after the resin has cured are left under water at reduced pressure so as to fully saturate the specimen (Figure 5/3(b)). After saturation the specimen is fitted into the lower half of a special cell, being sealed by O-Rings.

The cell is bolted together and filled with water, air being vented from the centre valve and the three-way tap turned so as to fill the calibrated capillary tube (see ISAT Method). By isolating the reservoir, movement of liquid along the capillary can be measured which corresponds to flow through the concrete specimen. It is well established that during the early stages of such a test the coefficient of permeability continues to decrease. Many explanations of this effect exist (10) but from the practical standpoint, measurements taken over a seven day period give results that can be usefully compared.

Plotting K_w against time allows performance comparisons to be made.

FUTURE TRENDS

Waterproofers, perhaps more than any other admixture, have developed in a "cut and try" way which has resulted in many products, few of which are proven and even less actually understood. It is likely that new products will evolve not as the result of scientific research but merely as a result of manufacturers extending and updating their present products. Therefore we have to look at the new polymer compositions to see whether they could have waterproofing potential.

The new melamine-formaldehyde resins appear to have remarkable plasticising properties which may alone improve waterproofness and, in addition, intensify the hydrophobic action of incorporated waterproofers such as silicones and metal soaps (15). Unfortunately these materials are relatively expensive.

Since damp-proofers appear to act on the surface only, there seems little point in distributing them throughout the bulk concrete. Perhaps compounds having preferential surface activity such that they might migrate to the concrete/air interface could be developed. Once at the surface these would oxidise or change so as to form a water repellent layer. This type of mechanism would allow expensive materials to be considered, since their function would be localised.

Any new product must be shown to be effective. To aid this a standard test pattern should be agreed, and it is a pity that integral waterproofers were not included in the draft British Standard on Admixtures.

One property of integral waterproofers which may have been overlooked is that of imparting "self-cleaning" properties to concrete containing them. Developments here could be beneficial, particularly in precast unit manufacture and systems building (11).

REFERENCES

1. J. Amer. Conc. Inst., 1971, No. 9, Proc. Vol. 68, pp. 646-676.
2. "Concrete", 1968, Vol. 2, No. 1, pp. 39-44.
3. J. Amer. Conc. Inst., 1950, Vol. 22, No. 1, pp. 25-52.
4. J. Amer. Conc. Inst., 1954, Vol. 26, No. 2, pp. 113-146.
5. J. Amer. Conc. Inst., 1963, No. 11, Proc. Vol. 60, pp. 1484-1524.
6. "The Chemistry of Cement and Concrete", F.M. Lea, pp. 602-604, Edward Arnold (Publishers) Limited, 1970.
7. "Concrete Technology and Practice", W.H. Taylor, pp. 190, 250, 392, Angus and Robertson, 1967.
8. T.C. Powers, L.E. Copeland and H.M. Mann, P.C.A. Journal, May 1959, Vol. 1, No. 2, pp. 38-48.
9. Mobil Product EU 80Y, now Mobilcer HM.
10. F. Kocataskin and E.G. Swenson, ASTM Bulletin, April 1958, pp. 67-72.
11. M. Levitt, "The ISAT — A Non-destructive Test for the Durability of Concrete", British Journal on Non-destructive Testing, Vol. 13, No. 4, 1970.
12. M. Vénuat, "Admixtures and the Treatment of Mortars and Concretes" Published by the author in Paris.
13. "Freezing and Thawing of Concrete and the Use of Silicones", Highway Research Record No. 18.
14. A.E. Desov, "The Effect of Silica-organic Compound GKZ-94 (Polyhydrosiloxane) on the Physico-technical Properties of Concrete and Mortars", Paper IV/7, International Symposium on Admixtures for Mortar and Concrete, Brussels, 1967.
15. A. Aignesberger and H.G. Rosenbauer, Tonindustrie Zeitung 1972, pp. 96, 29-34.

6
Mortar Plasticisers

H E Ackerman

REQUIREMENTS AND PROPERTIES OF MORTARS

Performance requirements for mortar

In the erection of brickwork, blockwork, masonry etc. mortar is required as a bedding to spread compressive loads, and as a bonding agent contributing some adhesion between components. These features impart stability to the structure. In addition, mortar seals the wall surface to prevent the ingress of wind and of wind-borne water in wet weather. The mortar must accommodate movement in the wall such as that due to thermal changes of dimension, or to the expansion of clay bricks or the contraction of concrete blocks or calcium silicate bricks during the early life of such materials. A mortar for brickwork or other walling must not possess excessive strength, but under building conditions it must develop the majority of its full strength fairly quickly.

Hydraulic lime mortars

Before the advent of Portland Cement mortars were made with hydraulic lime, that is lime which has cementing properties when mixed with water, and in which the necessary mortar properties were obtained as a result of the voids between the sand particles being filled with the smaller particles of the lime binder. The aggregate/binder ratio, in other words the sand/lime ratio, was about 3 : 1 in such mortars. There is still some small scale use of hydraulic lime mortar.

Renderings

Mortars are also used as a rendering over vertical surfaces of walls to impart decorative effect and weather resistance. Again, adequate strength, but not excessive strength, is required and Portland Cement is incorporated in order to achieve the required strength fairly quickly.

Working properties required

Of equal importance to the long term properties of mortars for brickwork and rendering, however, are their working properties. A good brickwork mortar must be easily mixed and should not segregate into its component parts whilst waiting to be used. It must work with the trowel satisfactorily, spread easily, and retain these properties during use. It must lose its water into porous bricks at an adequate but not excessive rate, otherwise there may be difficulty in adjusting the bricks into their final position in the wall. The mortar must develop good adhesion to the bricks so that when stressing subsequently occurs cracks do not open up which would allow ingress of wind and rain. Similar considerations prevail for rendering mortars.

Properties of unplasticised sand/cement

If, in the interests of achieving a rapid development of final strength, one were to make up a simple sand/cement mortar using the sands normally obtainable, adequate working properties could only be obtained if all the voids in the sand were filled with cement particles and water, and this would result in a sand/cement ratio of about 3 : 1. Such a mortar would be too strong for all except the densest bricks, would be expensive and would find use only in engineering work such as retaining walls. On the other hand, if just sufficient cement were used to give the required strength for normal exterior brickwork, say a 6 : 1 mortar, the resulting mix would normally be harsh and unworkable, would probably bleed water rapidly before it could be used, and would develop a poor bond between the components.

THE USE OF PLASTICISERS

Need for plasticiser

A material that we add to a sand/cement mortar of adequate mature strength, to impart the required working properties, we call a plasticiser.

There are available small quantities of building sands which by virtue of particle size distribution and particle shape enable mortars of adequate properties to be made without addition of a plasticiser. Such sands are rare in normal supply and are becoming rarer.

Non-hydraulic lime as a plasticiser

In addition to the use of hydraulic lime as the sole binder in lime/sand mortars, non-hydraulic lime has been in use as a plasticiser for cement mortars for many years, and imparts very good working properties. No early strength is imparted to the mortar by the non-setting lime; in fact, the strength is at first reduced. In the long term, a little extra strength is imparted to the set mortar by the lime because of the conversion of calcium hydroxide to calcium carbonate by absorption of carbon dioxide from the air.

Air-entraining plasticisers

However, the use of lime as a plasticiser is subject to certain disadvantages, and the main purpose of this chapter is to deal with an alternative method of imparting the necessary short term properties, by the addition of an air-entraining plasticiser. Use of such a material in place of lime is cheaper, and has at least one great advantage.

Definition

An air-entraining mortar plasticiser can be defined as an admixture which, when added to a sand/cement mix, causes the incorporation of air into that mix in the form of microscopic bubbles. These, together with the cement particles, fill the voids between the coarser sand particles so that the required working properties are obtained. On theoretical grounds one would expect such spherical bubbles of the right size to be the most efficient plasticising agents, because of their extra lubricating effect on the sand grains, and in practice this view is borne out by the fact that air-entrained mortars generally require less water to achieve a standard consistency.

Types of air-entraining plasticisers

We must turn now to the nature of air-entraining plasticisers. There are two main types. In the first, we have those whose active ingredient is a neutralised alkali-metal salt of Vinsol resin, a resin derived from wood. This is supplied either as the isolated salt in powder form, or as a solution in water, which gives greater convenience and reliability in use.

In the second type we have the synthetic surface active materials, typified by an alkyl benzene sulphonate. These are generally supplied as solids, for making into solution with water on site.

Masonry cement

Instead of using an admixture on site it is possible to use masonry cement, a Portland Cement in which an air-entraining agent together with a very small particle-sized mineral have been incorporated during manufacture. This frees the building operative from the task of adding the plasticiser, but it may involve the storage on site of more than one type of cement, since masonry cement could not be used for purposes where air entrainment would be a disadvantage, e.g. in reinforced concrete. Furthermore, there is less choice in the amount of plasticiser being used with various sands, since this is controlled by the cement content.

EFFECT OF PLASTICISERS

Air content achieved

Satisfactory plasticisation by air entrainment is normally obtained at air contents of between 15 and 22%. The air comes either from that initially present in the sand or cement or is gained from the surrounding air during the mixing process. These air bubbles persist in the mortar and on the setting of the cement component they are preserved in what might be described as encapsulated voids. Their presence is the cause of at least two important advantages which the air entrainment process has over the use of lime. (Lime, of course, gives rise to no air bubbles). These advantages are mainly concerned with the effects of frost.

Effect of frost on non-air-entrained mortar

We know that at the time of its use, and for some time afterwards, a non-air-entrained brickwork mortar can be considered to be a paste saturated, or almost saturated, with water. If freezing weather conditions develop at this time and no other precautions are taken to prevent the temperature of the newly laid brickwork from falling to freezing point nothing can stop the water in the mortar from freezing. At the same time as it changes to ice it increases in volume, causing disruptive pressures which separate from each other the partially bonded sand grains and stabilise them with ice in the extended location. Nothing can be observed to have happened until the mortar thaws out, at which time the solid components are not in proper contact with each other. The mortar hence exhibits little or no strength, crumbles to powder, and places the wall in danger of collapse.

Frost protection achieved by air entrainment

The incorporation of entrained air, however, gives rise to a vast number of evenly dispersed expansion chambers throughout the mass, and as some of the water freezes it is believed to push the remaining water into these, so preserving the interrelationship of the sand particles that existed before the frost. Freezing, therefore, does no damage, and on subsequent thawing the setting of the cement particles continues as if there had been no interruption, and full strength is developed. This is a major advantage of the use of an air-entraining plasticiser over the use of lime, which imparts no such frost protection.

It is perhaps necessary to reiterate that the freezing of water in a mortar in winter weather cannot be prevented by use of an "anti-freeze". The process used for the water in car radiators, for example, cannot be applied to the water in a mortar, since no practical freezing point depressant is known which does not irrevocably interfere with the application and setting process of the cement. Economically, too, the use of a true "anti-freeze" such as alcohol or glycol would be unacceptable, even if it were a technically viable process.

Official recommendations for air-entrainment

Official recommendations, such as those laid down in B.S. Codes of Practice and Building Research Establishment Digests, confirm the usefulness of an air-entraining plasticiser as a frost protective measure in winter building, in conjunction with other suitable mechanical precautions, and advise against the use of set-accelerating admixtures (which, as can be deduced from theoretical considerations, are largely ineffectual in practice).

Long-term durability of mortar

This frost-protective property imparted to a mortar by air entrainment is of long term importance, since the air bubbles remain capable of acting as expansion chambers throughout the life of the material.

Another advantage of air entrainment is that is imparts enhanced resistance to sulphate attack. BRE Digest No 160 "Mortars for bricklaying" reports that "...... there is evidence that aerated cement:sand mortars are more resistant to sulphate attack than cement:lime:sand mortars of the same strength designation".

Other effects of plasticisers

None of the plasticisers enters into any chemical reaction with the mortar or its constituents, and no colour change in the mortar is observable. The setting time and hardening time of the mortar are not altered, but there may be a change in the rate of loss of water into the surrounding brickwork because of the better control of "bleed" obtained on air entrainment.

Compared with an unplasticised mortar, air entrainment results in some loss of strength, because the same quantity of cement binder is being used to cement a larger volume of mortar. The difference is much less between aerated cement/sand mortar and an equivalent cement/lime/sand mortar, since they are more nearly equivalent in their cement contents per given volume of mortar.

As has been mentioned earlier, where a very strong mortar is required there is no suitable air-entrained mortar, since the reduction in strength due to the normal percentage of entrained air is unacceptable, and one cannot reliably obtain small proportions of air without recourse to frequent air volume measurement during preparation and use of the mortar.

Advantages of air-entraining plasticisers over lime

We have commented upon the frost protection advantage of air-entrained mortars over those plasticised with lime. We can now look at the additional advantages, at the mixing stage, to be gained from the use of a plasticising admixture. If we replace the lime in a 1 : 1 : 6 mix with the aforementioned typical Vinsol plasticiser we find that we are using about one gallon of plasticiser instead of 16 bags of slaked lime. Use of the latter, therefore, involves considerable extra dry storage space on site, and some inescapable mess caused by the handling of successive batches of dry, dusty material under open-air conditions. In contrast, one gallon of liquid may be stored almost anywhere, and introduced into gauging water as a simple stirring job. Furthermore, plasticising by hydrated lime is reliable only if the lime has been allowed to mature in the mix for several hours before the addition of cement, and under winter building conditions this is difficult and expensive to arrange without risk of the mix freezing during the conditioning process.

In addition to these savings in labour costs there is the simple saving in the material cost. Using the same figures as before, we must contrast the cost of about 1 gallon of Vinsol type plasticiser with that of 8 cwt of hydrated lime for an approximately equal output of prepared mortar.

SPECIFICATION AND MIXING

Typical specifications

It is perhaps useful to compare typical mix specifications for cement/lime/sand and air-entrained cement/sand mortars. This may be done by using the data in Table 6/1, which has been taken from the corresponding table in BRE Digest No 160.

Typical dosage rates

Dosage rates vary widely with the type of plasticiser and with the brand, but one typical Vinsol solution as supplied can be used at a rate of about

Table 6/1 Comparison of mortar mixes.

	Mortar group	Cement : lime : sand	Masonry-cement : sand	Cement : sand, with plasticiser	Mortar mixes (proportions by volume)
Increasing strength but decreasing ability to accommodate movements caused by settlement, shrinkage, etc	i	1 : 0 - ¼ : 3	–	–	
	ii	1 : ½ : 4 - 4½	1 : 2½ - 3½	1 : 3 - 4	
	iii	1 : 1 : 5 - 6	1 : 4 - 5	1 : 5 - 6	Where a range of sand contents is given, the larger quantity should be used for sand that is well graded and the smaller for coarse or uniformly fine sand.
	iv	1 : 2 : 8 - 9	1 : 5½ - 6½	1 : 7 - 8	
	v	1 : 3 : 10 - 12	1 : 6½ - 7	1 : 8	

Direction of changes in properties:
← equivalent strengths within each group →
→ increasing frost resistance
→ improving bond and resistance to rain penetration

Because damp sands bulk, the volume of damp sand used may need to be increased. For cement : lime : sand mixes, the error due to bulking is reduced if the mortar is prepared from lime : sand coarse stuff and cement in appropriate proportions; in these mixes 'lime' refers to non-hydraulic or semi-hydraulic lime and the proportions given are for lime putty. If hydrated lime is batched dry, the volume may be increased by up to 50 per cent to get adequate workability.

1 pint per 20 gallons (1 litre/160 litres) of gauging water, equivalent to about ½ pint per cwt (0.28 litre/50 kg) of cement in the mortar. A typical dosage rate for the detergent type is ¾ oz per cwt (21 g/50 kg) cement.

Mixing processes

If a liquid plasticiser is to be used the mixing process involves mixing the sand and cement together then gauging the mix with water containing the plasticiser. If the admixture is a solid, then it should be thoroughly incorporated with the dry sand and cement *before* gauging the mix with water. Satisfactory air entrainment can be achieved by hand mixing, or by using any of the conventional concrete mixers. An edge-runner, however, of the type at one time used for sand/lime mortars, would not be likely to entrain sufficient air.

Care should be taken not to over-mix, since with some plasticisers an excessive volume of air could be entrained, leading to a mortar of undesirable handling properties and undue weakness in the mature state.

In this respect it is worth noting that the Vinsol type of plasticiser gives mortars more resistant to excessive air entrainment than some of the synthetic surfactant type, which tend to require more care in use to guard against overdosing and overmixing errors.

SUMMARY

Finally, the advantages of using a mortar plasticiser compared with using lime may be summarised as follows:

Advantages of Air Entrainment over Lime in Mortar Plasticisation

1. Frost protection is afforded to mortar which is used in cold weather.
2. There is increased long term resistance to freeze/thaw cycles of mortar which becomes saturated.
3. Plasticised mortar has lower material cost.
4. Storage costs for raw materials are lessened.
5. An air-entraining plasticiser is easier and less messy to add to mortar, resulting in lower labour costs.
6. The mix does not need to mature before use.

BS for mortar plasticisers

Mortar plasticisers such as those discussed here are covered by BS 4887 : 1973 "Specification for Mortar Plasticisers".

Appendix

Concrete Admixtures – Data Sheets

NOTES

1. HEALTH AND SAFETY AT WORK ACT 1974:– advice on safety and handling of admixtures should be obtained by reference to the individual members of the Cement Admixtures Association.

2. BRITISH STANDARDS – BS 5075: Part 1: 1974 and BS 4887. Some admixtures are now covered by these British Standards and customers should approach individual members of the Cement Admixtures Association to obtain information on compliance.

KEY TO CHARTS

TYPE	C CONCRETE M MORTARS
ACTIVE INGREDIENTS	Formate (F) Stearate or Derivative (St) Naphthalene Sulphonate (NS) Polyvinyl Acetate (PV) Styrene Butadiene (SBR) Hydroxycarboxylic Acid (HC) Melamine Formaldehyde (MF) Polyhydroxy Compound (PHC) Chloride (Cl) Phosphates (Phos) Lignosulphonate (Lig) Surfactant (SA) Wood Resin Derivative (WR) Other (O)
MIN./MAX. DOSAGE (METRIC)	Typical Dosage per 50 kg cement — ml for liquids, g for powder. Where one value is given, this is the recommended dosage.
MIN./MAX. DOSAGE (IMPERIAL)	Typical Dosage per 50 kg cement — fl.oz. for liquids, oz. for powder. Where one value is given, this is the recommended dosage.
CHLORIDE AT MAXIMUM DOSAGE	Chloride added at maximum typical dosage — % anhydrous $CaCl_2$ on cement weight.
AIR ENTRAINMENT	
	Slight Moderate Considerable
	S M C
CONCRETE — Additional air entrainment	2% 2-4% 4%
— Total air entrainment	3% 3-6% 6%
MORTARS, etc. — Additional air entrainment	3% 3-15% 15%
— Total air entrainment	10% 10-22% 22%
ONE YEAR SHELF LIFE	Is shelf life more than 12 months under recommended storage conditions? YES/NO
EFFECT OF LOW TEMPERATURE	Under freezing conditions the product is A irretrievably damaged B useable after allowing to thaw and re-mixing C unaffected
SPECIFIC GRAVITY	Relative density of liquid.
CHLORIDE ION CONTENT	% Cl mass/mass (wt/wt). Note: $<$0.1% recorded as NIL.

Concrete admixtures (C references)

C1 Accelerating

BRAND NAME	MANUFACTURER	FORM	ACTIVE INGREDIENTS	MIN./MAX. DOSAGE (METRIC)	MIN./MAX. DOSAGE (IMPERIAL)	CHLORIDE AT MAX. DOSAGE	AIR ENTRAINMENT	ONE YEAR SHELF LIFE	AGITATION NEEDED BEFORE USE	EFFECT OF LOW TEMPERATURE	SPECIFIC GRAVITY	CHLORIDE ION CONTENT
ACCELERATOR	Blundell-Permaglaze	Liquid	Cl	2200	77½	1.4	S	YES	NO	B	1.23	16½
CEMFROPROOF	Cementation Chemicals	Liquid	Cl	600 / 1700	21 / 60	1.3	S	YES	NO	B	1.27	17½
COLEMANOID No. 9	Altro	Liquid	Cl	1500	53	1.4	S	YES	NO	B	1.34	22
TRICOSAL T4	Berk	Powder	F	500 / 3000	18 / 120	NIL	S	YES	NO	C	—	NIL
LILLINGTON No. 1 METALLIC LIQUID	Blundell-Permaglaze	Liquid	Cl/St	2200	77½	1.0	S	YES	YES	B	1.19	12
QUICKSOCRETE 'D'	C.B.P.	Liquid	Cl	1200 / 1700	40 / 60	1.4	S	YES	NO	B	1.32	20
SEALOSET	Sealocrete	Powder	O	1000	35	NIL	S	YES	NO	C	—	NIL
SOVEREIGN SUPER CONCENTRATED	Sovereign	Liquid	Cl	1000 / 1600	35 / 56½	1.5	S	YES	NO	B	1.23	25
STRIP-STRIKE	Hydrol	Liquid	Cl	750 / 1500	25 / 55	1.5	S	YES	NO	B	1.33	22
HY-PRO-CRAAM	Hydrol-Protex	Liquid	Cl	500	20	0.5	S	YES	NO	B	1.30	20
HY-ACE (Non-Chlor)	Hydrol-Protex	Liquid	F	500 / 3000	120	NIL	S	YES	NO	B	1.10	NIL
35/36% CALCIUM CHLORIDE LIQUOR	I.C.I.	Liquid	Cl	500 / 1550	18 / 54½	1.5	S	YES	NO	B	1.35	23
KYLJACK	Shell Composites	Liquid	Cl	2500	88	1.4	S	YES	NO	B	1.21	15
FEBCAST	FEB	Liquid	Cl	500 / 1750	18 / 62	1.5	S	YES	NO	B	1.34	20
FEBSPEED	FEB	Liquid	Cl	1500 / 3000	50 / 100	1.2	S	YES	NO	B	1.18	14
ANTIFROSTO	Sika	Liquid	Cl	1500	53	1.5	S	YES	NO	B	1.34	22

C2 Retarding

BRAND NAME	CORMIX P.2	RETARDER POWDER	SIKA RETARDER	TRICOSAL VZ 100	CORMIX R1
MANUFACTURER	Crosfield	C.B.P.	Sika	Berk	Crosfield
FORM	Liquid	Liquid Powder	Liquid	Liquid Powder	Liquid
ACTIVE INGREDIENTS	HC	Phos	PHC Phos	Phos	HC Phos
MIN./MAX. DOSAGE (METRIC)	280	100 500	100 1000	20 280	50 1000
MIN./MAX. DOSAGE (IMPERIAL)	10	4 16	3½ 35	2½ 70	1¾ 35
CHLORIDE AT MAX. DOSAGE	NIL	NIL	NIL	NIL	NIL
AIR ENTRAINMENT	S	S	S	S	S
ONE YEAR SHELF LIFE	YES	YES	YES	YES	YES
AGITATION NEEDED BEFORE USE	NO	NO	NO	NO	NO
EFFECT OF LOW TEMPERATURE	B	C	B	B	B
SPECIFIC GRAVITY	1.17	—	1.08	1.39	1.23
CHLORIDE ION CONTENT	NIL	NIL	NIL	NIL	NIL

C3 Normal water reducing

BRAND NAME	TRICOSAL BV	CONPLAST 211	CORMIX P.1	CORMIX P.3	CORMIX P.6	FLOCRETE N.	HY-PRO PACE	LANOMIX	SEALOPLAZ SPECIAL	SEALOPLAZ STANDARD	FEBFLOW STANDARD	PLASTIMENT – A40	PLASTIMENT – BV40	PLASTOCRETE – OC	FRIOPLAST – A3
MANUFACTURER	Berk	C.B.P.	Crosfield	Crosfield	Crosfield	Cementation Chemicals	Hydrol-Protex	Blundell-Permoglaze	Sealocrete	Sealocrete	FEB	Sika	Sika	Sika	Sika
FORM	Liquid	Liquid	Liquid	Liquid	Liquid	Liquid	Liquid	Liquid	Liquid	Liquid	Liquid	Liquid	Liquid	Liquid	Liquid
ACTIVE INGREDIENTS	Lig	Lig	Lig	PHC	Lig	Lig	Lig	Lig	HC	Lig	Lig	Lig	Lig	Lig	Lig SA
MIN./MAX. DOSAGE (METRIC)	100	140 210	140	140*	100 140	140	125*	140*	90 130	90 130	80 280	150 250	150 250	250	250*
MIN./MAX. DOSAGE (IMPERIAL)	3	5 7½	5	5*	3½ 5	5	4*	5*	3 4½	3 4½	3 10	5 9	5 9	9	9*
CHLORIDE AT MAX. DOSAGE	NIL	NIL	NIL	NIL	NIL	NIL	NIL	NIL	NIL	NIL	NIL	NIL	NIL	NIL	NIL
AIR ENTRAINMENT	S	S	S	S	S	S	S	S	S	S	S	S	S	S	M
ONE YEAR SHELF LIFE	YES	YES	YES	YES	YES	YES	YES	YES	YES	YES	YES	YES	YES	YES	YES
AGITATION NEEDED BEFORE USE	NO	NO	NO	NO	NO	NO	NO	NO	NO	NO	NO	NO	NO	NO	NO
EFFECT OF LOW TEMPERATURE	B	B	B	B	B	B	B	B	B	B	B	B	B	B	B
SPECIFIC GRAVITY	1.24	1.17	1.13	1.13	1.12	1.18	1.20	1.24	1.32	1.17	1.18	1.19	1.19	1.18	1.17
CHLORIDE ION CONTENT	NIL	NIL	NIL	NIL	NIL	NIL	NIL	NIL	NIL	NIL	NIL	NIL	NIL	NIL	NIL

Note: *Suggested rate of addition as starting point for trial mixes.

C4 Accelerating water reducing

BRAND NAME	TRICOSAL BVS	COLEMANOID NO. 1	CONPLAST NC	CONPLAST 'W'	FLOCRETE 'A'	PREMIX DOUBLE STRENGTH	SEALOFROST	CORMIX P.8	CONCRETE PLASTICISER & ACCELERATOR	FEBCAST SUPER	FEBFLOW ACCELERATING	FEBCAST P.3	FEBEXEL	FEBSPEED PLUS	SIKA-SET-CL	HY-PRO CRAAM
MANUFACTURER	Berk	Aitro	C.B.P.	C.B.P.	Cementation Chemicals	Sealocrete	Sealocrete	Crosfield	Shell Composites	FEB	FEB	FEB	FEB	FEB	Sika	Hydrol-Protex
FORM	Liquid	Liquid	Liquid	Liquid	Liquid	Liquid	Liquid	Liquid	Liquid	Liquid	Liquid	Liquid	Liquid	Liquid	Liquid	Liquid
ACTIVE INGREDIENTS	Lig O	Cl O	O	Cl Lig	Cl Lig	Cl Lig	Cl Lig	Cl HC	Cl Lig	Cl Lig	Cl Lig	Cl O	O WR	Cl WR	Cl PHC	Cl Lig
MIN./MAX. DOSAGE (METRIC)	80	1000 2000	1250	850 1700	850 1700	1250 2500	1250 2500	500 1500	1000 2000	500 1750	840 1680	900 1800	1000	1000	500 1700	500
MIN./MAX. DOSAGE (IMPERIAL)	3	35 70	44	30 60	30 60	44 88	44 88	18 53	35 70	18 62	30 60	32 64	35	35	18 60	20
CHLORIDE AT MAX. DOSAGE	NIL	1.1	NIL	1.4	1.4	1.5	1.5	1.4	1.5	1.45	1.5	1.5	NIL	1.5	1.5	0.5
AIR ENTRAINMENT	S	S	S	S	S	S	S	S	S	S	S	S	M	M	S	S
ONE YEAR SHELF LIFE	YES	YES	NO	YES	YES	YES	YES	YES	YES	YES	YES	YES	YES	YES	YES	YES
AGITATION NEEDED BEFORE USE	NO	NO	YES	YES	YES	NO	NO	NO	YES	NO	NO	NO	NO	NO	NO	NO
EFFECT OF LOW TEMPERATURE	B	B	B	B	B	B	B	B	B	B	B	B	C	C	B	B
SPECIFIC GRAVITY	1.24	1.23	1.24	1.31	1.33	1.26	1.26	1.33	1.27	1.33	1.33	1.33	—	—	1.33	1.30
CHLORIDE ION CONTENT	NIL	14	NIL	20	20	15	15	22	17	20	20	20	NIL	46	20	20

C5 Retarding water reducing

BRAND NAME	CONPLAST 'R'	CORMIX P.2	CORMIX P.5	FLOCRETE 'R'	RETARDO	SEALOPLAZ Retarding Grade	FEBFLOW Retarding	TRICOSAL VZ	PLASTIMENT VZ	FRIOPLAST – V3	SIKA RETARDER	SIKA RETARDER 2	HY-PRO PACE
MANUFACTURER	C.B.P.	Crosfield	Crosfield	Cementation Chemicals	Blundell-Permoglaze	Sealocrete	FEB	Berk	Sika	Sika	Sika	Sika	Hydrol-Protex
FORM	Liquid	Liquid	Liquid	Liquid	Liquid	Liquid	Liquid	Liquid	Liquid	Liquid	Liquid	Liquid	Liquid
ACTIVE INGREDIENTS	HC	HC	HC	HC	Lig	Lig PHC	Lig	Lig	PHC	PHC SA	PHC Phos	Phos PHC	Lig
MIN./MAX. DOSAGE (METRIC)	140 280	80 140	140*	150*	280*	90 130	150 300	70 280	100 250	100 250	100 1000	100 1000	125 —
MIN./MAX. DOSAGE (IMPERIAL)	5 10	3 5	7*	5½*	10*	3 4½	5 10	2.5 10	3.5 9	3.5 9	3.5 35	3.5 35	4
CHLORIDE AT MAX. DOSAGE	NIL	NIL	NIL	NIL	NIL	NIL	NIL	NIL	NIL	NIL	NIL	NIL	NIL
AIR ENTRAINMENT	S	S	M	S	S	S	S	S	S	M	S	S	S
ONE YEAR SHELF LIFE	YES	YES	YES	NO	YES	YES	YES	YES	YES	YES	YES	YES	YES
AGITATION NEEDED BEFORE USE	NO	NO	NO	NO	NO	NO	NO	NO	NO	NO	NO	NO	NO
EFFECT OF LOW TEMPERATURE	B	B	B	B	B	B	B	B	B	B	B	B	B
SPECIFIC GRAVITY	1.20	1.17	1.17	1.05	1.20	1.19	1.19	1.22	1.18	1.17	1.08	1.20	1.20
CHLORIDE ION CONTENT	NIL	NIL	NIL	NIL	NIL	NIL	NIL	NIL	NIL	NIL	NIL	NIL	NIL

Note: *Suggested rate of addition as starting point for trial mixes.

C6 Air entraining agents

BRAND NAME	MANUFACTURER	FORM	ACTIVE INGREDIENTS	MIN./MAX. DOSAGE (METRIC)	MIN./MAX. DOSAGE (IMPERIAL)	CHLORIDE AT MAX. DOSAGE	AIR ENTRAINMENT	ONE YEAR SHELF LIFE	AGITATION NEEDED BEFORE USE	EFFECT OF LOW TEMPERATURE	SPECIFIC GRAVITY	CHLORIDE ION CONTENT
SEALOPLAZ AEA	Sealocrete	Liquid	WR	30/60	1/2	NIL	M	YES	NO	B	1.04	NIL
AIREN	Blundell-Permoglaze	Liquid	WR	28*	1*	NIL	M	YES	NO	B	1.06	NIL
CEMAIRIN	Cementation Chemicals	Liquid	WR	28*	1*	NIL	M	YES	NO	B	1.05	NIL
CONPLAST A.E.A.	C.B.P.	Liquid	WR	30*	1*	NIL	M	YES	NO	B	1.02	NIL
CORMIX A.E.I.	Crosfield	Liquid	WR	45*	1½*	NIL	M	YES	NO	B	1.04	NIL
CORMIX P.5	Crosfield	Liquid	HC	140*	7*	NIL	M	YES	NO	B	1.17	NIL
CORMIX 83/20	Crosfield	Liquid	SA	200*	7*	NIL	C	YES	NO	B	1.04	NIL
SOVEX	Sovereign	Liquid	WR	60/120	2/4	NIL	M	YES	NO	B	1.03	NIL
HY-PRO AES	Hydrol-Protex	Liquid	WR	15*/30	½*/1	NIL	M	YES	NO	B	1.04	NIL
TRICOSAL LPS	Berk	Liquid	WR	20/60	0.6/1	NIL	M	YES	NO	B	1.06	NIL
FEBCRETE AEA	FEB	Liquid	WR	20/60	0.75/2	NIL	M	YES	NO	B	1.06	NIL
FRO-BE	Sika	Liquid	WR	25*	¾*	NIL	M/C	YES	NO	B	1.05	NIL
SIKA-AER	Sika	Liquid	SA	20*	¾*	NIL	M/C	YES	NO	B	1.02	NIL
FRIOPLAST – A.5	Sika	Liquid	Lig SA	250*	9*	NIL	M/C	YES	NO	B	1.17	NIL
CORMIX AE2	Crosfield	Liquid	SA	45*	1½*	NIL	M	YES	NO	B	1.02	NIL

Note: *Suggested rate of addition as starting point for trial mixes.

C7 Pumpability aids

BRAND NAME	CONPLAST 'PA'	CORMIX P.4.	PUMPING AID	FLOCRETE 'P'	FEBSLIP 200	TRICOSAL BVS	ACOSAL	HY-PRO PACE	SIKA PUMPCRETE	PLASTIMENT – L
MANUFACTURER	C.B.P.	Crosfield	Sealocrete	Cementation Chemicals	FEB	Berk	Berk	Hydrol-Protex	Sika	Sika
FORM	Liquid	Liquid	Liquid	Liquid	Liquid	Liquid	Liquid	Liquid	Liquid	Liquid
ACTIVE INGREDIENTS	Lig SA	HC	SA	Lig O	Lig O	Lig O	Lig O	Lig	O	Lig SA PHC
MIN./MAX. DOSAGE (METRIC)	140 210	140	200 400	500 1000	300 1250	80	350	125*	170	170
MIN./MAX. DOSAGE (IMPERIAL)	5 7½	5	7 14	17½ 35	10 40	3	10	4*	6	6
CHLORIDE AT MAX. DOSAGE	NIL	NIL	NIL	NIL	NIL	NIL	NIL	NIL	NIL	N L
AIR ENTRAINMENT	S	S	S	S	S	S	S	S	S	S
ONE YEAR SHELF LIFE	YES	YES	YES	YES	YES	YES	YES	YES	YES	YES
AGITATION NEEDED BEFORE USE	NO	NO	NO	NO	NO	NO	NO	NO	NO	NO
EFFECT OF LOW TEMPERATURE	B	B	B	B	B	B	B	B	B	B
SPECIFIC GRAVITY	1.16	1.08	1.04	1.01	1.06	1.24	1.18	1.20	1.18	1.18
CHLORIDE ION CONTENT	NIL	NIL	NIL	NIL	NIL	NIL	NIL	NIL	NIL	N L

Note: *Suggested rate of addition as starting point for trial mixes

C8 Integral waterproofers

BRAND NAME	MANUFACTURER	FORM	ACTIVE INGREDIENTS	MIN./MAX. DOSAGE (METRIC)	MIN./MAX. DOSAGE (IMPERIAL)	CHLORIDE AT MAX. DOSAGE	AIR ENTRAINMENT	ONE YEAR SHELF LIFE	AGITATION NEEDED BEFORE USE	EFFECT OF LOW TEMPERATURE	SPECIFIC GRAVITY	CHLORIDE ION CONTENT
COLEMANOID No. 1	Altro	Liquid	Cl O	1000 2000	35 70	1.1	S	YES	NO	B	1.23	14
CORMIX W.1	Crosfield	Powder	St	500	17½	NIL	S	YES	NO	C	—	NIL
CORMIX W.2	Crosfield	Liquid	St	750	26½	NIL	S	YES	NO	B	1.00	NIL
CORMIX W.3	Crosfield	Liquid	O	120 1000	4 35	NIL	S	YES	NO	B	0.92	NIL
HY-WAC	Hydrol	Liquid	St	1000	40	NIL	S	YES	YES	B	1.05	NIL
LILLINGTON No. 1 METALLIC LIQUID	Blundell-Permoglaze	Liquid	Cl St	2200	77½	1.0	S	YES	YES	B	1.19	12
MEDUSA	C.M.C.	Powder	St	1000	35	NIL	S	YES	NO	C	—	NIL
TRICOSAL SPERR	Berk	Liquid Powder	Lig St	180	6	NIL	S	YES	NO	B	1.15	NIL
PREMIX DOUBLE STRENGTH	Sealocrete	Liquid	Cl Lig	1250 2500	44 88	1.5	S	YES	NO	B	1.26	15
PROPALIN POWDER S	C.B.P.	Powder	St	500	18	NIL	S	YES	NO	C	—	NIL
PROPALIN 421	C.B.P.	Liquid	O HC	800 1000	28 35	NIL	S	YES	NO	B	1.09	NIL
SEALOPRUF SPECIAL	Sealocrete	Liquid	St	500	17½	NIL	S	YES	NO	B	1.01	NIL
SEALOPRUF STANDARD	Sealocrete	Liquid	Lig	500	17½	NIL	S	YES	NO	B	1.01	NIL
SETCRETE RMW	Cementation Chemicals	Liquid	Lig	250	9	NIL	S	YES	NO	B	1.14	NIL
SETCRETE 1	Cementation Chemicals	Liquid	O	1000	35	0.08	S	YES	YES	B	1.13	1.3
SOVEREIGN WATERPROOFER	Sovereign	Liquid	St	2000	70	NIL	S	YES	NO	B	1.04	NIL
INTEGRAL WATERPROOFER	Shell Composites	Paste	St	1000	35	NIL	S	YES	YES	B	1.13	<1
FEBPROOF RMC	FEB	Liquid	St Lig	500	18	NIL	S	YES	YES	B	1.04	NIL
FEBTITE LIQUID	FEB	Liquid	St	1000 2000	35 70	NIL	M	YES	NO	B	1.05	NIL
PLASTOCRETE NW	Sika	Liquid	Lig PHC	250	9	NIL	S	YES	NO	B	1.08	NIL
SIKA-1	Sika	Liquid	O	1000 1500	35 53	NIL	S	YES	YES	B	1.05	NIL
SIKALITE	Sika	Powder	St	1000	35	NIL	S	YES	NO	C	—	NIL

C9 Superplasticisers

BRAND NAME	CONPLAST M1	CORMIX SP1	MELMENT L 10	SEALOPLAZ SUPER	ACOSAL	SIKAMENT	SUPAFLO	ACOSAL NT
MANUFACTURER	C.B.P.	Crosfield	Hoechst	Sealocrete	Berk	Sika	Cementation Chemicals	Berk
FORM	Liquid	Liquid	Liquid	Liquid	Liquid	Liquid	Liquid	Liquid
ACTIVE INGREDIENTS	MF	NS	MF	SA Lig	Lig	NS	NS	Lig O
MIN./MAX. DOSAGE (METRIC)	650 2000	600 900	670 1340	500 1000	350	350 850	300 800	450
MIN./MAX. DOSAGE (IMPERIAL)	23 70	21 31	22 44	17½ 35	12.5	10 30	10 28	15.7
CHLORIDE AT MAX. DOSAGE	NIL	NIL	NIL	NIL	NIL	NIL	NIL	NIL
AIR ENTRAINMENT	S	S	S	S	S	S	S	S
ONE YEAR SHELF LIFE	YES	YES	YES	YES	YES	YES	YES	YES
AGITATION NEEDED BEFORE USE	NO	NO	NO	NO	NO	NO	NO	NO
EFFECT OF LOW TEMPERATURE	B	B	B	B	C	B	B	C
SPECIFIC GRAVITY	1.10	1.20	1.12	1.12	1.16	1.20	1.14	1.08
CHLORIDE ION CONTENT	NIL	NIL	NIL	NIL	NIL	NIL	NIL	NIL

Mortar admixtures (M references)

M1 Mortar plasticisers

BRAND NAME	CEBIX 112	CEBIX 112 Sachet	COLEMANOID SUPER 5	CORMIX MPI	CORMIX A.E.3	HY-PRO PLAM	PLASMOR	PLAZ	SOVEREIGN ADMIX	SOVEREIGN POWDERED MORTAR PLASTICISER	TROWELITE	MAXI STRENGTH	STANDARD MORTAR PLASTICISERS	SBD MORTAR PLASTICISERS	FEBMIX ADMIX	FEBMIX DH	SIKAMIX LIQUID	SIKAMIX POWDER
MANUFACTURER	C.B.P.	C.B.P.	Altro	Crosfield	Crosfield	Hydrol-Protex	Blundell-Permoglaze	Sealocrete	Sovereign	Sovereign	Cementation Chemicals	Shell Composites	Shell Composites	SBD Construction	FEB	FEB	Sika	Sika
FORM	Liquid	Sachet or Bottle	Liquid	Liquid	Liquid	Liquid	Liquid	Liquid	Liquid	Powder	Liquid	Liquid	Liquid	Liquid	Liquid	Powder	Liquid	Powder
ACTIVE INGREDIENTS	WR	WR	WR	SA	SA	WR	WR	WR	WR	Lig SA	WR	WR	WR	WR	WR	WR	WR	WR
MIN./MAX. DOSAGE (METRIC)	210 280	1 Sac or Bot	50	45*	50*	35 75	280	250	71 284	14 28	56 140	50	250	280	140 280	15 30	250	250
MIN./MAX. DOSAGE (IMPERIAL)	7½ 10	1 Sac or Bot	2	1½*	1½*	1.25 2.50	10	9	2½ 10	½ 1	2 5	2	9	10	5 10	0.5 1	9	9
CHLORIDE AT MAX. DOSAGE	NIL	NIL	NIL	NIL	NIL	NIL	NIL	NIL	NIL	NIL	NIL	NIL	NIL	NIL	NIL	NIL	NIL	NIL
AIR ENTRAINMENT	M	M	M	M	C	M	M	M	M	M	M	M	M	M	M	M	M	M
ONE YEAR SHELF LIFE	YES	YES	YES	YES	YES	YES	YES	YES	YES	YES	YES	YES	YES	YES	YES	YES	YES	YES
AGITATION NEEDED BEFORE USE	NO	NO	NO	NO	NO	NO	NO	NO	NO	NO	NO	YES	YES	NO	NO	NO	NO	NO
EFFECT OF LOW TEMPERATURE	B	B	B	B	B	B	B	B	B	C	B	B	B	B	B	C	B	B
SPECIFIC GRAVITY	1.01	1.04	1.11	1.02	1.01	1.09	1.02	1.03	1.02	–	1.05	1.09	1.02	1.02	1.02	–	1.02	–
CHLORIDE ION CONTENT	NIL	NIL	NIL	NIL	NIL	NIL	NIL	NIL	NIL	NIL	NIL	NIL	NIL	NIL	NIL	NIL	NIL	NIL

Note: *Suggested rate of addition as starting point for trial mixes.

M2 Integral waterproofers

BRAND NAME	COLEMANOID No. 1	CORMIX W.1	CORMIX W.2	CORMIX W.3	TRICOSAL SPERR	HY-WAC	LILLINGTON No. 1 METALLIC LIQUID	PROLAPIN '031'	SEALOREND	SETCRETE W.R.	SETCRETE 1	SOVEREIGN WATERPROOFER	INTEGRAL WATERPROOFER	FEBPROOF	FEBTITE POWDER	SIKA-1	SIKALITE
MANUFACTURER	Aitro	Crosfield	Crosfield	Crosfield	Berk	Hydrol	Blundell-Permoglaze	C.B.P.	Sealocrete	Cementation Chemicals	Cementation Chemicals	Sovereign Chemicals	Shell Composites	FEB	FEB	Sika	Sika
FORM	Liquid Powder	Powder	Liquid	Liquid	Liquid Powder	Liquid	Liquid	Liquid	Liquid	Liquid	Liquid	Liquid	Paste	Liquid	Powder	Liquid	Powder
ACTIVE INGREDIENTS	Cl O	St	St	O	Lig St	St	Cl St	St	St	St	St	St	St	St Lig	St WR	O	St
MIN./MAX. DOSAGE (METRIC)	1000 2000	500 g	750	120 1000	180	1000	– 2200	700 1300	600 1400	800	250	2000	1000	560	1000	– 1500	1000
MIN./MAX. DOSAGE (IMPERIAL)	35 70	17½	26½	4 35	6	40	77½	24 45	21 49½	28	9	70	35	20	35	– 53	35
CHLORIDE AT MAX. DOSAGE	1.1	NIL	NIL	NIL	NIL	NIL	1.0	NIL	NIL	NIL	0.16	NIL	NIL	NIL	NIL	NIL	NIL
AIR ENTRAINMENT	S	S	S	S	S	S	S	M	M	S	S	S	S	S	M	S	S
ONE YEAR SHELF LIFE	YES	YES	YES	YES	YES	YES	YES	YES	YES	YES	YES	YES	YES	YES	YES	YES	YES
AGITATION NEEDED BEFORE USE	NO	NO	NO	NO	NO	YES	YES	NO	YES	YES	NO	NO	YES	YES	NC	YES	NO
EFFECT OF LOW TEMPERATURE	B	C	B	B	B	B	B	B	B	B	B	B	B	B	C	B	C
SPECIFIC GRAVITY	1.23	–	1.00	0.92	1.15	1.05	1.19	1.05	0.98	1.01	1.13	1.04	1.13	1.14	–	1.05	–
CHLORIDE ION CONTENT	14	NIL	NIL	NIL	NIL	NIL	12	NIL	NIL	NIL	1.3	NIL	<1	NIL	NIL	NIL	NIL

M3 PVA and other resin admixtures

BRAND NAME	MANUFACTURER	FORM	ACTIVE INGREDIENTS	MIN./MAX. DOSAGE (METRIC)	MIN./MAX. DOSAGE (IMPERIAL)	CHLORIDE AT MAX. DOSAGE	AIR ENTRAINMENT	ONE YEAR SHELF LIFE	AGITATION NEEDED BEFORE USE	EFFECT OF LOW TEMPERATURE	SPECIFIC GRAVITY	CHLORIDE ION CONTENT
BLUE CIRCLE BONDING AGENT	C.M.C.	Liquid Paste	PV	9000	320	NIL	S	YES	YES	‡	1.09	NIL
CEBOND	C.B.P.	Liquid	PV	4000	140	NIL	S	YES	NO	‡	1.07	NIL
CEMPROVER	C.M.C.	Liquid	PV	23000	810	NIL	S	YES	YES	A	1.03	NIL
HY-BUC	Hydrol	Liquid	PV	6000	200	NIL	S	YES	NO	‡	1.07	NIL
SEALOBOND	Sealocrete	Liquid	PV	3000 12000	110 420	NIL	S	YES	NO	B	1.03	NIL
SEALOTAK	Sealocrete	Liquid	SBR	9000 12000	320 420	NIL	S	YES	NO	B	1.00	NIL
SOVEREIGN PVA	Sovereign	Liquid	PV	6000	210	NIL	S	YES	YES	‡	1.07	NIL
ECOSEAL	Berk	Liquid	PV	4000 12000	—	NIL	S	YES	YES	B	1.07	NIL
ISOBOND	Shell Composites	Liquid	PV	24000	845	NIL	S	YES	YES	‡	1.06	NIL
SBD SPEEDIBOND	SBD Construction	Liquid	PV	2500 5000	90 180	NIL	S	YES	YES	B	1.07	NIL
SBD MULSIBOND STANDARD	SBD Construction	Liquid	PV	3000 12000	40 400	NIL	S	YES	YES	B	1.01	NIL
FEBOND	FEB	Liquid	PV	4550 13600	160 450	NIL	S	YES	NO	‡	1.09	NIL
SIKAFIX	Sika	Liquid	PV	3000 15000	110 550	NIL	S	YES	NO	‡	1.08	NIL
SIKA LATEX	Sika	Liquid	SBR	† 18000	633	NIL	S	YES	NO	B	1.03	NIL
SIKADUR DISPERSION	Sika	Liquid	Epoxy	†	†	NIL	S	YES	YES	A	—	NIL

‡ Partially resistant, but protect from frost. † Dependent upon colour.

Concrete and mortar pigments

BRAND NAME	MORTEX	SEALANTONE CEMENT COLOURS	CONCRETE COLOURS	FEBTONE
MANUFACTURER	Sovereign	Sealocrete	Shell Composites	FEB
FORM	Liquid	Liquid	Powder	Powder
ACTIVE INGREDIENTS	Pig	Lig Iron Oxide	Iron Ox BR	Iron Ox BR
MIN./MAX. DOSAGE (METRIC)	9000	2500	1000 3000	1360 2720
MIN./MAX. DOSAGE (IMPERIAL)	320	88	35 105	48 144
CHLORIDE AT MAX. DOSAGE	NIL	NIL	NIL	NIL
AIR ENTRAINMENT	S	S	S	S
ONE YEAR SHELF LIFE	YES	YES	YES	YES
AGITATION NEEDED BEFORE USE	NO	YES	NO	NO
EFFECT OF LOW TEMPERATURE	B	B	C	C
SPECIFIC GRAVITY	1.08	—	—	—
CHLORIDE ION CONTENT	NIL	NIL	NIL	NIL

CONSTRUCTION LIBRARY,
Liverpool Polytechnic,
Clarence Street, L3 5TP.